PUBLIC HEALTH IN THE 21ST CENTURY

THE OCCUPATIONAL SAFETY AND HEALTH ACT: STANDARDS, PROVISIONS, RIGHTS

PUBLIC HEALTH IN THE 21ST CENTURY

Additional books in this series can be found on Nova's website under the Series tab.

Additional e-books in this series can be found on Nova's website under the e-book tab.

PUBLIC HEALTH IN THE 21ST CENTURY

THE OCCUPATIONAL SAFETY AND HEALTH ACT: STANDARDS, PROVISIONS, RIGHTS

CHRISTY KNIGHT
EDITOR

NOVINKA

New York

NOTICE TO THE READER

The Publisher has taken reasonable care in the preparation of this book, but makes no expressed or implied warranty of any kind and assumes no responsibility for any errors or omissions. No liability is assumed for incidental or consequential damages in connection with or arising out of information contained in this book. The Publisher shall not be liable for any special, consequential, or exemplary damages resulting, in whole or in part, from the readers' use of, or reliance upon, this material. Any parts of this book based on government reports are so indicated and copyright is claimed for those parts to the extent applicable to compilations of such works.

Independent verification should be sought for any data, advice or recommendations contained in this book. In addition, no responsibility is assumed by the publisher for any injury and/or damage to persons or property arising from any methods, products, instructions, ideas or otherwise contained in this publication.

This publication is designed to provide accurate and authoritative information with regard to the subject matter covered herein. It is sold with the clear understanding that the Publisher is not engaged in rendering legal or any other professional services. If legal or any other expert assistance is required, the services of a competent person should be sought. FROM A DECLARATION OF PARTICIPANTS JOINTLY ADOPTED BY A COMMITTEE OF THE AMERICAN BAR ASSOCIATION AND A COMMITTEE OF PUBLISHERS.

Additional color graphics may be available in the e-book version of this book.

Library of Congress Cataloging-in-Publication Data

ISBN: 978-1-63482-637-2

Published by Nova Science Publishers, Inc. † New York

CONTENTS

PREFACE

Through the Occupational Safety and Health Act of 1970 ("OSH Act" or "act"), Congress sought a nationwide approach to regulating workplace accidents and injuries. The act authorizes the Secretary of Labor to create and enforce workplace safety standards. Additionally, the act contains a "General Duty Clause," also enforced by the Secretary of Labor, which generally requires employers to provide workplaces that are free of potentially harmful hazards. The act created the Occupational Safety and Health Administration ("OSHA") and an Assistant Secretary of Labor for Occupational Safety and Health, to whom the Secretary of Labor has delegated his enforcement rights and obligations under the act. The act also established the Occupational Safety and Health Review Commission ("OSHRC"), an adjudicatory agency independent of the Department of Labor, and therefore OSHA, that is tasked with reviewing enforcement actions. OSHA enforces its standards and the General Duty Clause through inspections, citations, and penalties. Employers can seek review of OSHA enforcement actions first with OSHRC and then with U.S. Courts of Appeals. This book discusses, in depth, the standards, provisions and rights of the Occupational Safety and Health Act.

In: The Occupational Safety and Health Act
Editor: Christy Knight
ISBN: 978-1-63482-637-2

Chapter 1

THE OSH ACT: A LEGAL OVERVIEW*

Rodney M. Perry

SUMMARY

Through the Occupational Safety and Health Act of 1970 ("OSH Act" or "act"), Congress sought a nationwide approach to regulating workplace accidents and injuries. The act authorizes the Secretary of Labor to create and enforce workplace safety standards. Additionally, the act contains a "General Duty Clause," also enforced by the Secretary of Labor, which generally requires employers to provide workplaces that are free of potentially harmful hazards. The act created the Occupational Safety and Health Administration ("OSHA") and an Assistant Secretary of Labor for Occupational Safety and Health, to whom the Secretary of Labor has delegated his enforcement rights and obligations under the act. The act also established the Occupational Safety and Health Review Commission ("OSHRC"), an adjudicatory agency independent of the Department of Labor, and therefore OSHA, that is tasked with reviewing enforcement actions. OSHA enforces its standards and the General Duty Clause through inspections, citations, and penalties. Employers can seek review of OSHA enforcement actions first with OSHRC and then with U.S. Courts of Appeals.

---

* This is an edited, reformatted and augmented version of a Congressional Research Service publication R43768, prepared for Members and Committees of Congress, dated January 22, 2015.

Congress passed the Occupational Safety and Health Act of 1970[1] ("OSH Act" or "act") to "assure so far as possible every working man and woman in the Nation safe and healthful working conditions and to preserve our human resources ..."[2] In the years leading up to the OSH Act, approximately 14,500 workers died annually from work-related injuries, and the number of Americans dying from work-related injuries between 1966 and 1970 exceeded the number of Americans killed in the Vietnam War during that period.[3] Additionally, work-related injuries disabled approximately 2.2 million Americans each year.[4] Congress therefore found existing health and safety standards inadequate at protecting workers and passed the OSH Act, whereby it sought a "comprehensive, nationwide approach" to occupational health and safety.[5]

Background

The OSH Act obligated the Secretary of Labor to protect occupational safety and health as required by the act. However, the act also created the Assistant Secretary of Labor for Occupational Safety and Health,[6] and thus the Occupational Safety and Health Administration ("OSHA"), to whom the Secretary of Labor has delegated responsibility for his obligations under the act.[7] The OSH Act also created the Occupational Safety and Health Review Commission ("OSHRC")[8]—an adjudicatory agency that is independent of the Department of Labor and therefore OSHA—that is tasked with reviewing OSHA enforcement actions (i.e., citations and penalties).[9]

The OSH Act contains express limitations on its applicability. For example, Section 4(b)(1) of the act provides that the act does not apply "to working conditions of employees with respect to which other federal agencies ... exercise statutory authority" over occupational safety and health regulations and standards.[10] Congress apparently intended this language to prevent regulatory inefficiency by precluding OSHA from duplicating worker safety efforts in areas already regulated by other agencies.[11] To that end, health and safety regulations of other agencies that apply to a workplace hazard preempt OSHA's regulations regarding that hazard.[12] The OSH Act further expressly limits its applicability by allowing states to control occupational safety and health regulation through state plans.[13] If OSHA approves of a state plan, which requires, among other things, that the plan provide health and safety protection that is at least as adequate as the protections provided by OSHA,[14] then the state plan requirements, and not OSHA's standards and regulations, will apply.[15] Finally, the OSH Act applies only to employment performed in U.S. states and territories,[16] generally does not apply to federal, state, or local government employment,[17] and only applies to employers with employees (i.e., it does not apply to self-employed people).[18]

STANDARDS

The OSH Act contains two primary enforcement provisions, each of which furnishes a unique obligation upon employers. First, Section 5(a)(1) of the act, the so-called "General Duty Clause," requires all employers, regardless of industry, to provide workplaces that are free of potentially harmful hazards.[19] Second, the act mandates employer compliance with OSHA's workplace safety standards, some of which are industry specific.[20] The General Duty Clause and OSHA's workplace safety standards are treated the same for purposes of complaints, inspections, and citations. That is, employers are equally bound to comply with the General Duty Clause and applicable OSHA standards and both are similarly enforced through complaints, inspections, and citations. OSHA bears the burden of showing that an employer has violated the General Duty Clause or an OSHA standard by a preponderance of the evidence.[21] Additionally, employers cannot be held liable for violations of the General Duty Clause or an OSHA standard unless they had actual or constructive knowledge of the violation's existence.[22] Thus, after establishing that an employer has violated the General Duty Clause or an OSHA standard by a preponderance of the evidence, OSHA bears the added burden of proving that the employer did so with actual or constructive knowledge.[23]

General Duty Clause

Under the General Duty Clause, employers must provide employees with "employment and a place of employment which are free from recognized hazards that are causing or are likely to cause death or serious physical harm."[24] Apparently recognizing the impracticality of having OSHA issue safety standards for every type of employer and employment, Congress intended the General Duty Clause to be a catchall provision establishing a broad employer duty to protect worker health and safety in the absence of applicable OSHA-created standards.[25] To that end, when an OSHA-created standard applies to a hazard, an employer's compliance with the standard can be deemed to satisfy, and thereby preempt, the General Duty Clause.[26] However, this preemption occurs only when the employer successfully proves, as an affirmative defense, that the OSHA standard adequately addresses the hazard or, if the standard inadequately addresses the hazard, that the employer did not know of the inadequacy of the standard.[27] Thus, if an employer knows that a safety standard inadequately protects employees against a hazard, the employer must still comply with the General Duty Clause regarding that hazard.[28]

To successfully prove a violation of the General Duty Clause, OSHA must show that (1) a workplace condition poses a hazard to an employee; (2) the employer constructively or actually recognizes that the workplace condition is a hazard; (3) the hazard is causing, or is likely to cause, serious physical harm or death; and (4) feasible means exist to materially reduce or eliminate the hazard.[29] If OSHA successfully proves a violation of the General Duty Clause, it must then show that the employer violated the clause with actual or constructive knowledge for the employer to have violated the OSH Act.[30]

Hazard to an Employee

A General Duty Clause violation requires that a workplace condition pose a hazard to an employee. This does not require actual harm to an employee; the ability of a workplace condition to harm an employee can render it a hazard.[31] Additionally, OSHRC has recognized that a workplace condition can be a hazard absent a significant chance that the condition will injure an employee.[32] On the other hand, for a condition to be a hazard, it must occur in "other than a freakish or utterly implausible concurrence of circumstances."[33] Accordingly, for a workplace condition to be a hazard, it must be more than "utterly implausible" that the condition will injure or kill an employee, but there need not be a significant chance that it will injure or kill an employee.

"Recognized" Hazard

If a workplace condition is a hazard, OSHA must establish that it is a "recognized" hazard to prove a violation of the General Duty Clause. A recognized hazard is a workplace condition that is either recognized as hazardous by the employer's industry, or that the employer actually knows to be hazardous.[34] In determining that an industry recognizes a hazard, OSHRC has found persuasive evidence showing that, for example, accepted industry standards[35] or industry publications[36] acknowledge the hazard. Similarly, OSHA can establish an employer's actual knowledge of a hazardous condition by putting forth various types of evidence, including evidence showing that there were numerous previous accidents or injuries[37] or employee complaints.[38] An employer's actual knowledge of a hazardous condition can sustain a General Duty Clause violation even if the employer's industry has not recognized the condition as hazardous.[39]

Serious Harm or Death

If OSHA successfully establishes that a workplace condition is a recognized hazard, it must prove that the hazard is causing, or is likely to cause, serious physical injury or death to an employee. OSHRC has held that the probability that physical harm or death will occur, which it considers when determining whether a hazard exists as discussed above, is irrelevant in determining whether a recognized hazard is causing, or is likely to cause, serious physical harm or death to employees.[40] Rather, OSHRC considers the level of physical harm that an employee would face *if* an accident were to happen, no matter how slight the chance an accident would happen in the first place.[41] For example, in *Waldon Health Care Center*, a hazardous workplace condition presented only a small chance that employees would catch Hepatitis B, a virus from which, OSHRC noted, most people fully recover.[42] Even so, OSHRC found serious physical harm or death likely because a small percentage of people that catch Hepatitis B die or do not fully recover from it, and the hazardous condition therefore had the chance, though apparently slight, of seriously injuring or killing employees.[43] Previous incidents wherein the recognized hazard caused serious physical harm or death are not *necessary* to establish that the hazard is causing, or is likely to cause, injury or death.[44] However, it appears that previous injuries or deaths are *sufficient*, unless rebutted, to establish that the hazard is causing, or is likely to cause, serious physical injury or death.[45] Reviewing courts generally defer to OSHRC determinations that a recognized hazard is likely to cause serious injury or death.[46]

Feasible Means Exist to Reduce or Eliminate Hazard

Courts and the OSHRC generally agree that Congress did not intend the General Duty Clause to protect employees against all conceivable workplace hazards.[47] Rather, Congress intended the clause to protect employees against only hazards that are *preventable*.[48] Accordingly, if OSHA successfully establishes that a workplace condition poses a recognized hazard that is likely to cause serious physical harm or death to an employee, OSHA must then prove that the hazard is preventable; that is, feasible means of materially reducing the hazard exist.[49] In practice, litigation of whether feasible means of reducing a recognized hazard exist appears to unfold in two steps. First, because OSHA bears the burden of establishing that a feasible means of abatement exists, OSHA must initially show that an employer could feasibly take specific actions that would materially reduce a risk to employees.[50] Next, the employer often responds by arguing that OSHA's suggested actions are infeasible.[51]

OSHA bears the burden of showing that a feasible means of materially reducing a hazard exists, which requires it to show that specific abatement actions by an employer would feasibly and materially reduce the hazard.[52] These abatement actions generally appear to fall into two categories: physical means of abatement[53] and abatement by implementing new or additional procedures.[54] When proposing an abatement action, OSHA can show its feasibility through evidence showing, for example, that similar employers in the industry have successfully implemented the means of abatement[55] or industry standards recommend the means of abatement.[56] If an employer does not rebut OSHA's evidence by putting forth his own evidence showing that the suggested means of abatement would be infeasible or ineffective at reducing the hazard, OSHA has successfully met its burden of establishing that a feasible means exists to reduce or eliminate a hazard.[57]

Employers often attempt to rebut OSHA's evidence by showing that OSHA's suggested means of abatement is infeasible or ineffective. They may do so by, among other things, proving that the suggested means of abatement is cost prohibitive[58] or, though possible to implement and its implementation would significantly and materially reduce the cited hazard, the suggested means of abatement causes some greater harm.[59] For example, in *Cargill, Inc.*, OSHA suggested that a grain elevator and feed mill operator upwardly extend sheet casings at the top of a vertical conveyor through the roof of a building so that, in the event of a grain dust explosion, air could escape and prevent secondary explosions and fires.[60] In response, the employer presented evidence that extending the casings would pierce structural beams which support the roof, thereby seriously weakening the roof, and OSHA did not challenge the employer's evidence on this point.[61] OSHRC, accepting the employer's evidence, found OSHA's suggested means of abatement infeasible because, while it may have reduced the risks of secondary grain dust explosions, it would have created a new risk by weakening the roof such that the roof could collapse.[62]

Actual or Constructive Knowledge

In addition to the four elements discussed above, to successfully hold an employer in violation of the General Duty Clause, OSHA must prove that an employer had actual or constructive knowledge of the condition that violated the OSH Act.[63] This requirement of actual or constructive knowledge on the part of the employer also applies to violations of OSHA-created standards, discussed below.[64] OSHRC has held that OSHA does not have to prove that an employer knew the requirements of the act or knew that a condition violated the act to establish an employer's knowledge.[65] Rather, OSHA must show that the employer knew of the physical conditions giving rise to the violation.[66]

Proving actual knowledge is generally as straightforward as it sounds: OSHA can do so through evidence showing that the employer knew that the violating condition existed.[67] Conversely, demonstrating an employer's constructive knowledge of a violating condition is less straightforward. This requires OSHA to prove that, though the employer did not know the condition existed, the employer could have known about the condition through the exercise of reasonable diligence.[68] OSHA can prove constructive knowledge by showing, for example, that the employer did not have an adequate safety program for promoting compliance with safety standards[69] or that the condition was in plain view.[70] The amount of time that a violating condition existed is relevant to determining whether it could have been discovered through the exercise of reasonable diligence; the longer a condition existed, the more likely courts and OSHRC are to find that the condition could have been discovered through reasonable diligence.[71] For example, in *R.P. Carbone Construction Co.*, a construction company allegedly failed to exercise reasonable diligence to protect its employees from potential falls while building a steel structure.[72] OSHRC found, and the U.S. Court of Appeals for the Sixth Circuit upheld, that because employees' lack of fall protection was plainly observable throughout the construction site and existed for approximately two weeks, the exercise of reasonable diligence could have informed the employer that the employees lacked fall protection.[73]

A supervisory employee's actual or constructive knowledge of a condition can generally be imputed to the supervisor's employer.[74] A supervisory employee is one to whom an employer has delegated authority over other employees, even if only temporarily.[75] In the U.S. Court of Appeals for the Fifth Circuit ("Fifth Circuit"), however, if a supervisor's own conduct caused a condition that violates the act, the supervisor's knowledge of the condition cannot be imputed to his employer unless his conduct was foreseeable.[76] The Fifth Circuit has held that a supervisor's violative conduct is not foreseeable, and therefore a supervisor's knowledge of his own violative conduct cannot be imputed onto an employer, if an employer has sufficient safety policies, training, and disciplinary procedures in place.[77]

OSHA-Created Standards

In addition to the obligations under the General Duty Clause, employers must "comply with occupational safety and health standards" that OSHA promulgates pursuant to Section 5(a)(2) of the OSH Act.[78] The OSH Act defines occupational safety and health standards as standards requiring "conditions, or the adoption or use of one or more practices, means, methods, operations, or processes, reasonably necessary or appropriate to provide safe or healthful employment and places of employment."[79] OSHA-created standards generally take two forms: specific standards and general standards. Specific standards narrowly apply to specified industries or activities (e.g., a standard applicable to logging).[80] General standards apply to classes of hazards that may be found throughout a number of industries (e.g., fire prevention standards).[81] In order to establish that an employer violated an OSHA-created standard, OSHA bears the burden of proving that (1) the standard applies to the cited workplace condition; (2) the employer did not comply with the terms of the standard; and (3) employees had access to the cited workplace condition.[82] Additionally, as is also necessary to establish a violation of the General Duty Clause as discussed above, for OSHA to prove a violation of OSHA-created standards it must prove that the employer had actual or constructive knowledge of the cited condition.[83] More specifically, OSHA must establish that the employer knew, or could have known through reasonable diligence, of the existence of the cited condition.[84]

In determining whether a standard applies to an employer's conduct, OSHRC generally considers the activity that the employer was engaged in when the citation arose. For example, in *Pico Industries*, OSHRC observed that when an employer was "engaged in steel erection activity," a standard applicable to steel erection applied.[85] Under OSHA regulations, if multiple standards apply to an activity or hazard, the most specific standard preempts more general standard(s).[86] The reverse is also true: general standards apply when there is no applicable specific industry standard.[87]

OSHA standards include, for example, those found in Code of Federal Regulations (CFR) Title 29 Parts 1910 (general standards), 1915 (shipyard standards), 1917 (marine terminal standards), 1918 (longshoring standards),

1926 (construction standards), and 1928 (agriculture standards). Employers are responsible for determining which standards apply to them and following the requirements of applicable standards even in the absence of any action by OSHA.[88] That is, even if OSHA has received no complaints about an employer, has not inspected an employer, and has not issued a citation to an employer, the employer must comply with standards applicable to them.[89] However, employers obtain exemptions from standards through variances.

Variances

The OSH Act permits employers to apply for exemptions from the requirements of OSHA-created standards or regulations under specified circumstances. Generally, OSHA must approve such exemptions, called variances, before an employer can permissibly fail to comply with an OSHA standard or the employee is in violation of the standard.[90] In other words, variances excuse only failure to comply with OSHA standards occurring after OSHA grants a variance; they do not retroactively remedy violations of standards.[91] The OSH Act authorizes OSHA to grant four types of variance: (1) temporary variances; (2) permanent variances; (3) experimental variances; and (4) variances required for national defense.[92]

Temporary Variances

The OSH Act allows employers to apply for, and OSHA to grant, temporary variances from OSHA-created standards when an employer cannot comply with a standard by the time the standard goes into effect.[93] To obtain such a variance, an employer must show three things. First, the employer must show that he cannot comply with a standard by its effective date because professional or technical personnel, materials, or equipment needed to comply with the standard are unavailable, or because the employer cannot finish necessary facility alterations or construction by the effective date.[94] Second, the employer must show that he is protecting his employees against the hazard covered by the standard despite not following the standard.[95] Third, the employer must show that a program is in place for complying with the standard as quickly as practicably possible.[96] OSHA cannot grant a temporary variance unless the employer's employees are first given notice of, and the opportunity to participate in, a hearing on the variance.[97] Temporary variances expire at the earlier of (1) the date the employer comes into compliance with the standard or (2) one year.[98] However, OSHA can grant renewal of a temporary variance up to two times.[99]

Permanent Variances

In addition to temporary variances, the OSH Act permits employers to apply for, and OSHA to grant, permanent variances from OSHA-created standards.[100] However, OSHA can only grant such permanent variances under narrowly prescribed circumstances.[101] More specifically, a permanent variance requires an employer to show, by a preponderance of the evidence, that his "conditions, practices, means, methods, operations, or processes" will provide the same level of health and safety protection as compliance with the standard.[102] In granting a permanent variance, OSHA must determine, on the record after the opportunity for inspection and, where appropriate, hearing, that the employer has successfully met his burden.[103] Additionally, before OSHA can grant a permanent variance, affected employees must receive notice of, and the opportunity to participate in, a hearing on the application for a permanent variance.[104]

Permanent variances are permanent in the sense that they generally do not contain an expiration date.[105] However, OSHA can modify or revoke a permanent variance at any time six months after its approval.[106] An employer, employees, or OSHA through its own motion can apply to modify or revoke a permanent variance.[107]

Experimental Variances

The OSH Act allows OSHA to grant variances when doing so is necessary for an employer to participate in an experiment that demonstrates the effectiveness of novel workplace health and safety techniques.[108] Either OSHA or the Secretary of Health and Human Services must approve these experiments.[109]

Variances Required for National Defense

Under the OSH Act, OSHA can grant variances from standards "as ... necessary and proper to avoid serious impairment of the national defense."[110] OSHA must determine that a variance is necessary for national defense on the record after notice and opportunity for hearing.[111] National defense variances last no more than six months, but employers can apply to renew them.[112]

Employer Recordkeeping, Reporting, and Posting Requirements

The OSH Act and OSHA regulations require certain employers to maintain and distribute information. More specifically, some employers must keep records of workplace injuries and illnesses, some of which they are then required to report to OSHA. The OSH Act and OSHA regulations further require employers to disseminate certain information to employees via posting requirements.

Recordkeeping

The OSH Act gives OSHA fairly broad authority to require employers to keep records of workplace injuries and illnesses.[113] The act authorizes OSHA to promulgate regulations requiring employers to maintain records of, and report, non-minor injuries, deaths, and illnesses that are work-related.[114] Under this authority, OSHA has passed generally applicable regulations requiring certain employers to maintain workplace injury and illness records and submit such records to OSHA.[115] OSHA's recordkeeping requirements for workplace injuries and illnesses are perhaps best considered by reference to two questions. First, as a threshold matter, to what employers do OSHA's recordkeeping regulations for workplace injuries and illnesses apply? Second, if OSHA's recordkeeping regulations apply to an employer, what workplace injuries and illnesses must be recorded and reported?

OSHA's recordkeeping requirements for workplace injuries and illnesses generally apply to all employers within the OSH Act's scope, subject to two broad exceptions.[116] First, the recordkeeping requirements do not apply to employers that had a total of 10 or fewer employees *at all times* during the last calendar year.[117] Second, the recordkeeping requirements do not apply to employers falling within one of the "low hazard retail, service, finance, insurance, or real estate" employer groups listed in the relevant OSHA regulations.[118] This list of exempted employer groups is extensive and specific; it includes, for example, hardware stores, gasoline service stations, used car dealers, and apparel and accessory stores.[119]

The recordkeeping regulations require employers operating multiple establishments, or physical locations where business is conducted,[120] to distinguish between exempted and non-exempted establishments for recordkeeping purposes.[121] For example, if an employer owns a factory that produces clothes (i.e., an establishment that is not on the list of exempt employer groups) and an apparel store that sells the clothes (i.e., an establishment that is on the list of exempt employer groups), the employer would have to maintain workplace injury and illness records for the factory but not the apparel store, provided that the employer has a total of 11 or more employees. Even if an employer qualifies for an exception from recordkeeping requirements, he must still maintain records if selected to participate in OSHA's annual employer survey, discussed below in the context of reporting requirements, or asked to do so by the Bureau of Labor Statistics (BLS).[122]

If an employer is required to keep workplace injury and illness records, he does not need to keep records on all workplace injuries and illnesses.[123] Rather, the employer is only required to keep records on injuries and illnesses that are (1) "work-related," (2) "new case[s]," and (3) qualify as injuries and illnesses as defined in the relevant regulations.[124] An injury or illness is work-related if an event in the workplace caused, contributed to, or significantly aggravated the injury or illness.[125] If an injury or illness results from workplace events, it is generally presumed to be a work-related injury.[126] Additionally, covered employers are only required to keep records of "new cases," meaning the employee has not had a previous injury or illness recorded of the same type affecting the same body part or, if the employee has, he recovered completely in the interim between the previous injury or illness and the current one.[127] An employee has recovered completely if all signs or symptoms have disappeared.[128] Finally, a qualifying injury or illness is generally one resulting in "death, days away from work, restricted work or transfer to another job, medical treatment beyond first aid, loss of consciousness ... [or] significant injury or illness diagnosed by a physician or other licensed healthcare professional."[129]

Reporting

Under the OSH Act and OSHA regulations, employers are generally required to report information on workplace injuries and illnesses to OSHA outside of an inspection[130] in two circumstances. First, employers falling within the OSH Act's coverage, whether required to keep records or not, must report serious work-related injuries to OSHA.[131] Such serious work-related injuries include those resulting in: (1) deaths of employees; (2) hospitalizations that lead to any employee receiving in-patient treatment; or (3) an employee's amputations or loss of an eye.[132] Employers must report work-related deaths to OSHA within eight hours of their occurrence,[133] but have 24 hours to report work-related in-patient hospitalizations, amputations, or eye losses.[134] Reports of serious injuries must be made to OSHA via telephone, or through electronic submission at OSHA.gov.[135]

The second instance wherein employers must report workplace injury and illness data to OSHA outside of the context of an inspection occurs when OSHA selects an employer to participate in its annual survey.[136] OSHA regulations require the randomly selected employers in certain industries that receive OSHA's annual survey form to complete the form and send it back to OSHA.[137] The form generally requires employers to submit information from their injury and illness records.[138] However, as discussed earlier in this report, not all employers are required to keep injury and illness records under OSHA regulations. If OSHA sends an annual survey form to an employer that is exempted from the injury and illness recordkeeping requirements, the employer is required to keep such records for the following year and subsequently return the survey form to OSHA.[139]

In November 2013, OSHA proposed a rule that would eliminate OSHA's annual survey of employers, but heighten employers' obligations to report injury and illness data to OSHA.[140] Under the rule, employers that are already required to keep injury and illness records would have to submit information from these records to OSHA electronically.[141] The rule thus does not appear to impose additional recordkeeping obligations on employers. The frequency of the compulsory submissions required by the rule would vary based on the size of the employer.[142] Employers that had 250 or more employees in the previous calendar year would be required to submit injury and illness information four times per year.[143] Employers that had 20 or more employees in the previous calendar year and are in certain designated industries would have to submit the required data every year.[144] Additionally, all employers, regardless of their size, would have to submit injury and illness information to OSHA if OSHA informs them that they must do so.[145] Once the information is submitted, OSHA would make it publicly available, minus any data whose public release is precluded by the Freedom of Information Act, the Privacy Act, or any specific provisions of OSHA regulations.[146]

Posting

The OSH Act and OSHA regulations contain requirements apparently aimed at ensuring employers communicate to their employees about employee rights and obligations under the act. These requirements mandate that employers post certain materials at worksites where employees can readily see them. Posting requirements under the OSH Act and OSHA regulations direct employers to (1) display the OSHA poster at each work site;[147] (2) post annual summaries of injury illness and data records if they are required to maintain such records;[148] (3) post any citations that they receive;[149] and (4) post petitions for modifications of abatement dates subsequent to facing OSHA citations.[150]

The OSH Act mandates that OSHA promulgate regulations requiring employers to "keep their employees informed of their protections and obligations under [the act], including the provisions of applicable standards" through "posting of notices or other appropriate means."[151] Pursuant to this mandate, OSHA regulations require covered employers to post a notice, furnished by OSHA, at their worksites.[152] This notice, known colloquially as the "OSHA poster," informs employees that, for example, they have the right to notify employers about workplace hazards, they can request OSHA inspections if they believe there are hazardous conditions at their workplace, and they have the right to see OSHA citations issued to their employer.[153] Employers must display the OSHA poster in "conspicuous place[s] or places where notices to employees are customarily posted."[154] Employers must also make sure the OSHA poster remains unaltered and undefaced.[155]

Covered employers must also post annual summaries of injury and illness records if they are required to keep such records.[156] Like the OSHA poster, employers must post such annual summaries in "conspicuous place[s] or places where notices to employees are customarily placed."[157] Further, like the OSHA poster, employers must ensure that the annual summary posting is not altered or defaced.[158]

In addition to the OSHA poster and annual summaries of injury and illness records, OSHA regulations require employers to immediately post any citations, or unedited copies thereof, that they receive.[159] A citation must generally be posted where the alleged violation giving rise to the citation occurred.[160] However, if it is not practicable to post the citation where the alleged violation occurred, an employer can post the citation in a "prominent place" where all affected employees can readily observe it.[161] Citations must remain posted until the latter of the date on which the employer eliminates the cited violation or three working days.[162] Contesting a citation does not absolve an employer of his obligation to post the citation.[163] However, under such circumstances the employer can also post a notice indicating that he is contesting the citation, including his reasons for doing so.[164]

If an employer petitions for modification of the abatement date of a citation, OSHA regulations require the employer to post the petition. When OSHA issues a citation to an employer for violating a workplace safety standard, such citations contain an "abatement date," or a date by which the employer must come into compliance with the cited safety standard.[165] The employer can petition OSHA to modify the abatement date when, despite good faith efforts, he has been unable to come into compliance with requirements of a citation by the abatement date because of factors that are "beyond his reasonable control."[166] The employer must post a copy of such a petition in a conspicuous place that would provide notice to all affected employees, or near the location giving rise to the violation.[167] The petition must remain posted for 10 working days.[168]

OSHA INSPECTIONS

The OSH Act grants OSHA the authority to enter and inspect sites wherein employees perform work for an employer.[169] The OSH Act further permits OSHA to inspect all relevant structures, equipment, and conditions found within such worksites.[170] However, OSHA's authority to enter and inspect worksites is limited in two ways. First, any OSHA entrance into, and inspection of, a worksite must be "reasonable."[171] More specifically, OSHA can only enter and inspect workplaces at "reasonable times" such as during work hours, "within reasonable limits," and "in a reasonable manner."[172] Second, the Supreme Court has held that OSHA inspections are subject to the Fourth Amendment, and thus OSHA needs a warrant to inspect worksites without employer permission.[173]

Inspection unreasonableness is an affirmative defense to a citation resulting from that inspection that must be raised and proven by an employer.[174] To succeed, an employer must point toward unreasonable OSHA conduct in violation of the OSH Act.[175] The employer must also show that OSHA's unreasonable conduct prejudiced the employer.[176] The inquiry into whether an OSHA inspection was unreasonable is factual, varying from case-to-case. For example, OSHRC has found an OSHA inspection unreasonable when a city building inspector gave an employer 10 days to correct violations of building codes and forwarded the employer's building code violations to OSHA, which then conducted an inspection before the 10 day correction period expired.[177] After inspection, OSHA cited the employer for the same hazards as the city building inspector, which OSHRC deemed "inherently unfair and unreasonable."[178] If an employer succeeds in establishing that an inspection was unreasonable in violation of the OSH Act, any evidence obtained through that inspection cannot be used against the employer.[179]

In addition to being limited by the OSH Act reasonableness restrictions, the Supreme Court held in *Sec'y of Labor v. Barlow's* that OSHA inspections are limited by the Fourth Amendment, and OSHA therefore needs a warrant to inspect an employer's worksite over the employer's objections.[180] In *Barlow's*, when an OSHA inspector went to the defendant's electrical and plumbing installation business to conduct an inspection, the defendant refused the inspector entry citing his Fourth Amendment rights.[181] The Court held that the Fourth Amendment prohibition against unreasonable governmental searches applies to both homes and workplaces.[182] The court observed that, generally, warrantless governmental searches are presumptively unreasonable under the Fourth Amendment.[183] Accordingly, despite noting that nothing in the language of the OSH Act appears to require a warrant or any other process before an OSHA inspection,[184] OSHA inspectors generally cannot conduct an inspection without first obtaining a warrant.[185] However, this warrant requirement does not apply to OSHA inspections that are conducted with an employer's consent.[186]

If proper authority for an inspection exists (i.e., the inspection is reasonable under the OSH Act and complies with the warrant requirement if the employer objects to inspection), OSHA can conduct both programmed and unprogrammed inspections.

Types of Inspections

OSHA generally conducts two types of inspections: unprogrammed inspections and programmed inspections. Unprogrammed inspections are initiated in response to a specific incident or event. In contrast, programmed inspections are those routine inspections that are scheduled under OSHA's objective administrative procedures.

Unprogrammed Inspections

Unprogrammed inspections, or inspections that OSHA initiates in response to a specific incident or event, often take one of two forms. First, unprogrammed inspections occur in response to employee complaints or third party referrals submitted to OSHA.[187] Second, unprogrammed inspections occur after OSHA receives a report of a serious accident.[188]

The OSH Act allows employees to request an OSHA inspection if they believe that an imminent danger or a violation of a health or safety standard poses a physical threat to employees.[189] The act requires that any such employee request for an inspection be in writing, detail the grounds for the inspection with "reasonable particularity," and be signed by the employee or employee's representative.[190] Once an employee requests an OSHA inspection, OSHA determines whether there are "reasonable grounds" to believe that the alleged imminent danger or violation exists.[191] If OSHA determines that such reasonable grounds are present, OSHA must conduct an inspection.[192] If OSHA determines, however, that reasonable grounds do not exist, it must inform the requesting employee of such and is not obligated to conduct an inspection.[193] Though not addressed in the OSH Act, OSHA regulations allow an employee to internally appeal OSHA's determination that no reasonable grounds for an inspection exist.[194] If, on review, OSHA again finds inspection unwarranted, such a finding is "final and not subject to further review."[195] OSHA similarly conducts inspections based on referrals, which are distinguished from employee complaints because they generally originate from third parties.[196] OSHA may conduct inspections based on referrals via, for example, information reports by media sources made directly to OSHA or disseminated through the news.[197] OSHA may also conduct inspections in response to reports by other federal, state, or local government agencies.[198]

Unprogrammed OSHA inspections can also result from the reports of serious injuries that employers are required to make to OSHA. As discussed earlier in this report, all employers subject to the OSH Act must report to OSHA any workplace accidents resulting in an employee's death, or resulting in the in-patient hospitalization of three or more employees.[199] OSHA policies require that all such deaths or injuries be "thoroughly investigated" via inspection to determine whether a violation of the OSH Act or OSHA standards contributed to the death or injury.[200]

Programmed Inspections

In addition to unprogrammed inspections, OSHA conducts programmed inspections, or routine inspections that are scheduled under the objective administrative procedures of an inspection program or plan.[201] OSHA conducts multiple types of programmed inspection. For example, OSHA conducts programmed inspections for specific industries, such as parts of the maritime and construction industries.[202] OSHA also maintains an inspection program, known as the Site-Specific Targeting Program, for worksites that have 40 or more employees and have high rates of worker injuries and illnesses.[203]

Inspection Process

Under the OSH Act, any person who gives an employer unauthorized advance notice of an inspection can face up to a $1,000 fine and six months in prison.[204] OSHA regulations clarify when advance notice of an inspection is authorized, and thus can be given without violating the act. Under these regulations, advance notice is only permitted when (1) there is an imminent danger, so that the employer can attempt to remedy the danger immediately; (2) an inspection would be best conducted after normal business hours or other special preparations must be completed before an inspection; (3) it is necessary to ensure that parties who must be allowed to accompany an inspector can be present; and (4) OSHA determines advance notice would better effectuate inspection.[205]

OSHA is authorized to set the time for an inspection, subject to the reasonableness restrictions discussed earlier in this report.[206] Even so, OSHA generally conducts inspections during regular working hours unless "special circumstances indicate otherwise."[207] When an OSHA inspector arrives at a worksite for an inspection, he must present his credentials to the employer and "explain the nature and purpose of the inspection."[208] Further, if the employer must keep injury and illness records, the OSHA inspector can ask for copies of such records, which the employer must provide within four hours.[209] At the onset of the inspection, the OSHA inspector must inform the employer of what records he wants to review.[210]

The OSH Act allows both employers and employees to have a representative accompany an OSHA inspector during an inspection.[211] OSHA regulations further allow additional employer or employee representatives to accompany OSHA inspectors when inspectors find that additional representatives will assist inspection.[212] Employees' representatives must generally also be employees of the employer.[213] However, inspectors have the discretion to permit third parties to serve as employee representatives if an inspector determines that such representation is "reasonably necessary" to conducting an "effective and thorough" worksite inspection.[214] Additionally, inspectors can deny any representative from accompanying them during inspections when the representative's accompaniment would inhibit the inspection.[215]

During a worksite inspection, OSHA inspectors can collect environmental samples, take pictures or videos, or use any other reasonable investigation methods.[216] OSHA inspectors may also consult with employees regarding potential health and safety concerns, and employees must be given the opportunity to inform the inspector of any OSH Act violations that they believe exist at the worksite.[217] If an inspector uncovers an imminent danger that could immediately cause death or serious injury, he must inform affected employees and employers.[218] The inspector will then encourage the employer to remedy the danger.[219] If the employer refuses to or cannot remedy the danger, the OSH Act grants federal district courts jurisdiction to restrain any workplace conditions or practices posing such imminent dangers.[220]

A regulatory requirement that inspections not unreasonably disrupt workplace operations constrains OSHA inspector conduct during an inspection.[221] Additionally, the OSH Act generally requires inspectors to keep confidential any information obtained during an inspection containing, or with the potential to reveal, trade secrets.[222] However, inspectors can disclose such information in two circumstances: (1) when needed to effectuate enforcement of the act, they can disclose such information to officers and employees and (2) they can disclose such information in proceedings under the act.[223] When an inspector discloses information containing trade secrets during a proceeding, the entity hearing the proceeding (i.e., OSHA, the OSHRC, or a court) issues any orders that are appropriate to protect the information from further disclosure.[224]

At the end of an inspection, the OSHA inspector conducts a closing conference with the employer and employee representatives.[225] This conference can occur jointly—with both representatives— or separately, on-site or via telephone.[226] At this closing conference, the inspector discusses anything found during the inspection that appears to violate the act or raise any other relevant issues.[227] The OSHA inspector also informs the employer and employee representatives of their rights and obligations moving forward and answers any of their questions.[228]

CITATIONS AND PENALTIES

If an inspection shows that an employer has violated the OSH Act or an OSHA standard, OSHA issues the employer citations and, potentially, penalties. The act prescribes the procedural and substantive requirements for citations and penalties. The act further provides that penalties vary depending on the type of violation.

After an inspection, OSHA reviews the report of its inspector to determine whether the inspection uncovered a violation of the OSH Act or an OSHA standard.[229] If, after reviewing the inspection report, OSHA believes that an employer has violated the OSH Act or an OSHA standard, the act requires OSHA to issue a citation to the employer.[230] OSHA must issue this citation "with reasonable promptness" no more than six months after the OSH Act or OSHA standard violation.[231] The OSH Act requires that citations be in writing, particularly describe the nature of the violation, provide notice of which provision of the OSH Act or OSHA standard the employer allegedly violated, and contain an abatement date, or date by which the employer must remedy the violation.[232]

OSHA may issue a notice of proposed penalty within a reasonable time of an inspection after, or concurrent with, a citation.[233] OSHA must send any notice of proposed penalty to an employer by personal service or certified mail.[234] Additionally, a notice of proposed penalty must inform the employer that the proposed penalties will become final unless the employer seeks their appeal within 15 working days of receiving the notification.[235] Under the act, penalties vary depending on the seriousness and type of violation that occurred. Violations fall into one of three levels of seriousness: (1) serious violations; (2) non-serious violations; and (3) de minimis violations. Additionally, violations at each level of seriousness can result in increased penalties if the violation is willful, results from a failure to abate harm, or is part of an employer's repeated violation of the act.

Penalties for Serious Violations

A violation of the OSH Act or OSHA standards is serious if there is a "substantial probability" that the violation could result in death or serious physical harm for an employee.[236] Courts and OSHRC have held that this requirement does not mean OSHA must establish that there is a substantial probability that an accident *will* occur.[237] Rather, to establish a serious violation of the act, OSHA must show that there is a substantial probability that if an accident were to occur, no matter its chances of happening, it could result in death or serious physical harm in "other than freakish or utterly implausible" circumstances.[238] Even if a violation could result in death or serious physical harm for an employee, the violation will not be considered serious if the employer did not, and could not through the exercise of reasonable diligence, know of the violation.[239] If an employer commits a serious violation, the OSH Act requires OSHA to issue the employer a penalty no greater than $7,000.[240]

Penalties for Non-Serious Violations

The OSH Act classifies "non-serious" violations distinctly from "serious" violations, and provides different penalties for the two.[241] However, unlike serious violations, the act does not define non-serious violations other than to provide that they are violations "determined not to be of a serious nature."[242] OSHRC has interpreted this language to include violations that have a "direct and immediate relationship" with occupational health and safety, but do not present a substantial probability that death or serious physical harm will result.[243] Unlike serious violations, for which OSHA *must* issue penalties up to $7,000, OSHA *may* issue penalties up to $7,000 for non-serious violations.[244]

Penalties for De Minimis Violations

De minimis violations are those that do not have a direct and immediate relationship with safety and health.[245] Such violations occur when, for example, an employer complies with a proposed standard rather than the standard in effect and the proposed standard provides equal or greater employee protection, or follows the intent of a standard but deviates slightly from its specific requirements in a way that poses no direct or immediate threat to employee safety or health.[246] De minimis violations do not result in citations; if an employer commits a de minimis violation, OSHA issues the employer a notice instead of a citation.[247] Notices of de minimis violations are not subject to the same requirements as citations (e.g., they need not contain an abatement date[248] and there is no requirement that they be issued within six months of the alleged violation).[249] De minimis violations cannot result in penalties.[250]

Penalties for Willful Violations

The OSH Act permits OSHA to issue increased penalties to "any employer who willfully ... violates the requirements of [the act]."[251] However, the act does not define a willful violation. Courts and the OSHRC have held that a willful violation is one that an employer commits with either intentional disregard of the requirements of the act, or plain indifference for employee safety.[252] An employer intentionally disregards the requirements of the act when he actually knows what the act requires yet fails to comply with the act.[253] Even if an employer does not actually know of the requirements of the act, he can be found to have willfully violated the act if his behavior shows plain indifference toward employee safety, which requires OSHA to show that the employer's state of mind was such that even if he were aware of the violation, he would not care.[254] Employers have not willfully violated the act if they reasonably, and in good faith, believed their conduct was not in violation of the act.[255] Additionally, it is generally harder for OSHA to prove a willful violation of the General Duty Clause than a willful violation of an OSHA standard because it is more difficult to prove that an employer intentionally disregarded, or was plainly indifferent to, "the broad duty to 'furnish a place of employment free from recognized hazards.'"[256]

An employer that willfully violates the OSH Act faces a mandatory penalty of between $5,000 and $70,000.[257] Further, if an employer's willful violation of an OSHA standard causes an employee's death, the employer shall, upon conviction, be punished by a fine of up to $10,000 or imprisonment for up to six months.[258] If the employer had previously been convicted of such a willful violation causing death, the employer can be punished by a fine of up to $20,000 or imprisonment for up to one year.[259]

Penalties for Failure-to-Abate Violations

OSHA can issue a Notification of Failure to Abate if an inspection reveals that an employer failed to bring a violating condition, for which OSHA has issued the employer a citation, into compliance with the OSH Act by the abatement date within the citation.[260] For a Notification of Failure to Abate to withstand review, OSHA must prove that the original citation became a final OSHA order,[261] and that the violative conditions for which OSHA originally cited the employer still existed upon re-inspection.[262] If OSHA successfully meets its burden, an employer can potentially raise a number of defenses. For example, the employer can argue that the original citation did not particularly describe the violation such that he lacked notice of what was needed to abate the hazard,[263] or that the original cited hazard, which became final due to operation of law rather than adjudication, did not violate the OSH Act.[264] Under the OSH Act, if OSHA issues a Notification of Failure to Abate to an employer, the employer can face a penalty of up to $7,000 for each day past the abatement date that the employer fails to correct the hazard.[265]

Penalties for Repeated Violations

Unlike a failure-to-abate violation, which results from an employer's failure to remedy an *existing* violation that they have already been cited for, a repeated violation results from a *new* violation.[266] This new violation must be "substantially similar" to a previously cited violation which resulted in a final order against the employer.[267] To meet its burden of establishing that the latter violation is substantially similar to the prior violation, OSHA can show that both citations result from the employer's failure to comply with the same standard.[268] The burden then shifts to the employer to show that the two violations are not substantially similar by, for example, showing that they involve different conditions or hazards.[269] Violations of the same specific standards, rather than general standards, make two violations more likely to be substantially similar because the violations are more likely to involve the same conditions and hazards.[270] When two violations result from different standards, OSHA can generally prove that they are substantially similar by showing that the violations resulted from substantially similar hazards.[271] OSHA can issue penalties of up to $70,000 for repeated violations.[272]

Process of Adjudicating Citations and Penalties

Employers have 15 working days upon receiving a citation or notice of proposed penalty to notify OSHA that they intend to contest the notice or citation, which can include protest of the abatement date contained within the citation.[273] If the employer does not notify OSHA of his intent to contest the citation or notice of proposed penalty within this 15-working-day window, the citation and notice of proposed penalty are deemed final orders of OSHRC and are not subject to agency or judicial review.[274] Employees also have a right to protest, though it is narrower than an employer's right to protest: they can protest only the abatement date of a citation on the basis that the allotted abatement time is unreasonable.[275] This limited employee right to protest is subject to the same 15-working-day time constraint as employers' right to review.[276] When an employer or employee notifies OSHA of its intent to protest a citation or notice of proposed penalty, OSHA "immediately advise[s]" OSHRC of such notification.[277] OSHRC then commences its independent review process.[278]

A protest before OSHRC first goes to an OSHRC Administrative Law Judge (ALJ), who hears it.[279] The ALJ then issues a written decision.[280] Any party adversely affected by the ALJ's decision can then petition to OSHRC for further review.[281] Additionally, an OSHRC Commissioner may direct review of an ALJ's decision on his own motion.[282] In either circumstance, no party has a right to OSHRC review of an ALJ's decision; review is discretionary with OSHRC.[283] If OSHRC does not order review of an ALJ's decision within 30 days of the docketing date of the decision, the decision becomes a final order of OSHRC.[284] If OSHRC does review an ALJ's decision, it then issues a final order on the matter.[285]

Any person adversely affected by a final OSHRC order can seek its appeal in (1) the U.S. Court of Appeals wherein the violation allegedly occurred; (2) the U.S. Court of Appeals wherein the employer has its principal office; or (3) the U.S. Court of Appeals for the D.C. Circuit.[286] Additionally, OSHA can seek appeal of an OSHRC order.[287] Petition for review of an OSHRC order, whether by an adversely affected party or OSHA, must be filed within 60 days of the OSHRC order.[288] Filing a petition for review of an OSHRC order does not stay the order; the order remains in effect until a court of appeals vacates it.[289] On review, the courts of appeal find OSHRC's findings of facts conclusive if the facts are supported by substantial evidence.[290]

End Notes

[1] P.L. 91-596, 84 Stat. 1590 (1970).

[2] 29 U.S.C. §651(b).

[3] S.Rept. 91-1282, at 1 (1970). For comparison, in the most recent year for which data is available, 2012, the U.S. Bureau of Labor Statistics reported 4,628 occupational fatalities. U.S. BUREAU OF LABOR STATISTICS, REVISION TO THE 2012 CENSUS OF FATAL OCCUPATIONAL INJURIES (CFOI) COUNTS 1 (2014), available at http://www.bls.gov/iif/oshwc/cfoi/cfoi_revised12.pdf.

[4] Id.

[5] Id.

[6] See Occupational Safety and Health Act of 1970, P.L. 91-596, §29-30, 84 Stat. 1590, 1618-19 (codified as amended at 29 U.S.C. §553, 5 U.S.C. §5315, and 5 U.S.C. §5108).

[7] Delegation of Authority and Assignment of Responsibility to the Assistant Secretary for Occupational Safety and Health, 77 Fed. Reg. 3,912, 3912 (Jan. 25, 2012).

[8] See 29 U.S.C. §661.

[9] 29 U.S.C. §659(c).

[10] 29 U.S.C. §653(b)(1). The act is also inapplicable to workplace conditions that are subject to the safety regulations of state agencies responsible for nuclear materials under §274 of the Atomic Energy Act of 1954. Id.

[11] Organized Migrants in Cmty. Action v. Brennan, 520 F.2d 1161, 1167 (D.C. Cir. 1975) (finding the "clearly enunciated legislative intent which is manifest in section 4(b)(1)" is "giving the Secretary omnibus authority to regulate occupational safety and health, ... [while] avoid[ing] the wasteful duplication that would result where another federal agency was also providing for the occupational safety of a class of workers"); Pennsuco Cement and Aggregates, Inc., 8 BNA OSHC 1378, 1383 (No. 15462, 1980).

[12] AMR Servs. Co., Inc., No. 89-1764 at *1 (OSHRC Aug. 9, 1990).

[13] See 29 U.S.C. §667.

[14] 29 U.S.C. §667(c)(2).

[15] 29 U.S.C. §667(e). At the time of writing, OSHA has approved the plans of 21 states and Puerto Rico. 29 C.F.R. part. 1952.

[16] 29 U.S.C. §653(a).

[17] 29 U.S.C. §652(5) (defining "employer" in the context of the OSH Act as "a person engaged in a business affecting commerce who has employees, but ... not the United States (not including the United States Postal Service) or any State or Political subdivision of state."). The act requires federal agencies to establish and maintain their own workplace safety programs. 29 U.S.C. §668. Additionally, the act requires that state plans apply to state and local employment in states that have chosen to adapt state plans. 29 U.S.C. §667(c)(6).

[18] See 29 U.S.C. §652(5).

[19] See 29 U.S.C. §654(a)(1).

[20] See 29 U.S.C. §654(a)(2).

[21] See JPC Grp., Inc., 22 BNA OSHC 1859, 1860 (No. 05-1907, 2009); see also Otis Elevator Co., 21 BNA OSHC 2204, 2205 (No. 03-1344, 2007).

[22] See Henry J. Kaiser Co., 12 BNA OSHC 1008, 1009 (No. 83-1277, 1984).

[23] See id (applying the actual or constructive knowledge requirement in the context of a violation of an OSHA-created standard); see also Z and S Constr. Co., Inc. 77 BNA OSHC 1728, 1731 (No. 78-1908, 1979) (applying the actual or constructive knowledge requirement in the context of a General Duty Clause violation).

[24] 29 U.S.C. §651(a)(1).

[25] S.Rept. 91-1282, at 9-10 (1970) ("The committee recognizes that precise standards to cover every conceivable situation will not always exist. This legislation would be seriously deficient if any employee were killed or seriously injured on the job simply because there was no specific standard applicable to a recognized hazard which could result in such a misfortune. Therefore, to cover such circumstances, the committee has included a requirement to the effect that employers are to furnish employment and places of employment which are free from recognized hazards to the health and safety of their employees.").

[26] See 29 C.F.R. §1910.5(f) ("An employer who is in compliance with any standard in this part shall be deemed to be in compliance with the requirement of section 5(a)(1) of the Act ....").

OSHA is not required to make a preliminary showing that a specific standard does not apply when bringing action for violation of the General Duty Clause. See Safeway v. OSHRC, 382 F.3d 1189, 1194 (10th Cir. 2004). Rather, courts generally treat applicable specific standards as affirmative defenses to a General Duty Clause citation. See id. Thus, the burden is on employers to show that the General Duty Clause does not apply because the cited hazard is governed by a specific standard. See id.

[27] UAW v. Gen. Dynamics Land Sys. Div., 815 F.2d 1570, 1577 (D.C. Cir. 1987).

[28] Id; Safeway, Inc. v. OSHA, 382 F.3d 1189, 1194 (10th Cir. 2004).

[29] SeaWorld of Fla. v. Perez, 748 F.3d 1202, 1207 (D.C. Cir. 2014); Nelson Tree Servs., Inc. v. OSHRC, 60 F.3d 1207, 1209 (6th Cir. 1995); Roberts Sand Co., v. Sec'y of Labor, No. 13-11478, 2014 WL 2566932, at *1, *1 (11th Cir. June 9, 2014). Erickson Air-Crane, Inc., No. 07-0645 at *2 (OSHRC Mar. 2, 2012).

[30] See Z and S Constr. Co., Inc. 77 BNA OSHC 1728, 1731 (No. 78-1908, 1979); see also Otis Elevator Co., 21 BNA OSHC 2204, 2208 (No. 03-1344, 2007).

[31] See Titanium Metals Corp. of Am. v. Usery, 579 F.2d 536, 542 (9th Cir. 1978) (noting that "it is beyond dispute that an accident need not occur for a violation of [the General Duty Clause] to properly be found ..."); see also Nat'l Realty and Construction Co., Inc. v. OSHRC, 489 F.2d 1257, 1267 (D.C. Cir. 1973) ("To establish a violation of the general duty clause, hazardous conduct need not actually have occurred...").

[32] Valley Interior Sys., Inc., 21 BNA OSHC 2224, 2227-8 (No. 06-1395, 2007).

[33] Id (citing Waldon Health Care Ctr., 16 BNA OSHC 1052, 1063 (No. 89-2804, 1993)).

[34] Dura Division, 7 BNA OSHC 1938, 1939 (No. 78-3748, 1979); Fabi Const. Co. v. Sec'y of Labor, 508 F.3d 1077, 1081 (D.C. Cir. 2007).

[35] See Coleco Industries, Inc., 14 BNA OSHC 1961, 1967 (No. 84-546, 1991) (observing that OSHA's reliance on an industry standard to show industry recognition of hazards is permitted).

[36] See Culberson Well Serv., Inc., 11 BNA OSHC 1677, 1680 (No. 82-1162, 1983) (finding that industry safety manuals were "evidence of industry recognition of the hazard" at issue).

[37] Carlyle Compressor Co. v. OSHRC, 683 F.2d 673, 676 (2nd Cir. 1982) (observing that an employer had actual knowledge of a hazardous condition in part because the employer knew of prior incidents wherein the condition injured an employee); Wal-Mart Stores, Inc., No. 09-1013 at *30 (OSHRC Apr. 5, 2011) (observing, in the context of hazard recognition, that "actual knowledge of a hazard may be gained by means of prior accidents, prior injuries, employee complaints, and warnings communicated to the employer by an employee.").

[38] Wal-Mart Stores, Inc., No. 09-1013 at *30 (OSHRC Apr. 5, 2011) (observing, in the context of hazard recognition, that "actual knowledge of a hazard may be gained by means of prior accidents, prior injuries, employee complaints, and warnings communicated to the employer by an employee."); see St. Joe Minerals Corp. v. OSHRC, 647 F.2d 840, 845 (8th Cir. 1981) (finding that a hazard was recognized through an employer's actual knowledge in part because an employee had brought the hazard to the employer's attention).

[39] See Brennan v. OSHRC, 494 F.2d 460, 464 (8th Cir. 1974).

[40] Waldon Health Care Ctr., 16 BNA OSHC 1052, 1062 (No. 89-2804, 1993) (noting that OSHRC "... has made clear, the criteria for determining whether a hazard is 'causing or likely to cause death or serious physical harm' is not the likelihood of an accident or injury, but whether, if an accident occurs, the results are likely to cause death or serious harm."); Brian Hanley Logging, 20 BNA OSHC 1228, 1233 (No. 02-1152, 2003).

[41] See Walden Health Care Ctr., 16 BNA OSHC at 1062.

[42] Id.

[43] Id.

[44] See Titanium Metals Corp. of Am. v. Usery, 579 F.2d 536, 542 (9th Cir. 1978); see also Am. Phoenix, Inc., No. 11- 2969 at *14 (OSHRC Mar. 13, 2014) (finding serious injury or death by fire likely though no employees were previously injured by fire and the employer's citation stemmed from an inspection).

[45] See, e.g., Valley Interior Sys., Inc., 21 BNA OSHC 2224, 2230 (No. 06-1395, 2007) (noting that where an employee died after a malfunctioning aerial lift collapsed, the likelihood of death or serious injury was undisputed); Hudspeth-McDowell Enterprises, Inc., 6 BNA OSHC 1900, 1907 (No. 77-3133, 1978) (holding that prior serious employee injuries are sufficient evidence that a recognized hazard is causing serious physical injury or death to employees).

[46] See, e.g., Nat'l Realty and Constr. Co. v. OSHRC, 489 F.2d 1257, 1265 at n.33 (D.C. Cir. 1973) ("... the Commission's determination of likelihood [of serious physical harm] should be accorded considerable deference by the courts."); Kelly Springfield Tire Co., Inc. v. Donovan, 729 F.2d 317, 323 (5th Cir. 1984) ("The likelihood of the accident itself has received only low level scrutiny, leaving the Commission's expertise predominant in the area."); Illinois Power Co. v. OSHRC, 632 F.2d 25, 29 (7th Cir. 1980).

[47] See Nat'l Realty and Constr. Co., Inc., v. OSHRC, 489 F.2d 1257, 1266 (D.C. Cir. 1973) (observing that through the General Duty Clause, "Congress intended to require elimination only of preventable hazards"); see also Champlin Petroleum Co. v. OSHRC, 593 F.2d 637, 640 (5th Cir. 1979) (noting that "[t]he general duty obligation ... is not designed to impose absolute liability or respondeat superior liability for employees' negligence. Rather, it requires the employer to eliminate only 'feasibly preventable' hazards"); Metric Constructors, Inc., 16 BNA OSHC 1698, 1700 (No. 92-3322, 1994).

[48] Nat'l Realty and Constr. Co., Inc., 489 F.2d at 1266; Champlin Petroleum Co., 593 F.2d at 640.

[49] Nat'l Realty and Constr. Co., Inc., 489 F.2d at 1267; Roberts Sand Co., v. Sec'y of Labor, No. 13-11478, 2014 WL 2566932, at *1, *1 (11th Cir. June 9, 2014).

[50] See Nat'l Realty and Constr. Co., Inc., 489 F.2d at 1267. If OSHA fails to provide evidence describing what actions an employer could feasibly have taken to abate a hazard and the effectiveness of those actions, it has failed to meet its burden of proof, and a General Duty Clause violation will not stand. See id. at 1268.

[51] See, e.g., Performance Site Mgmt., 19 BNA OSHC 2054, 2057-59 (No. 01-0956, 2002) (wherein an employer argued that OSHA's recommended means of abatement, implementing new safety procedures, would not have materially reduced a recognized hazard because the suggested procedures were no different from the employer's existing procedures in practice).

[52] Pelron Corp., 12 BNA OSHC 1833, 1836 (No. 82-388, 1986) ("[OSHA] must show specific additional measures [employer] should have taken ... that would have materially reduced the risk of harm ...").

[53] See, e.g., Cargill, Inc., 10 BNA OSHC 1398, 1401 (No. 78-5707, 1982) (wherein OSHA's suggested means of abatement of a risk of explosions and fires was to physically move equipment to increase ventilation and decrease air pressure).

[54] For example, OSHA has argued that employers improperly used generalized warnings urging employees to avoid hazards or act in a safe manner when more specific safety precautions would feasibly, significantly, and materially reduce a hazard. See Alabama Power Co., 13 BNA OSHC 1240, 1243 (No. 84-357, 1987). OSHRC rejected OSHA's argument in

Alabama Power Co. after concluding that the employer's caution to employees was "not a mere generalized caution to employees to work safely." Id. at 1244.

[55] Cont'l Oil Co. v. OSHRC, 630 F.2d 446, 449 (6th Cir. 1980).

[56] General Dynamics Land Sys. Div., 15 BNA OSHC 1275, 1288 (No. 83-1293, 1991).

[57] United States Steel Corp., 12 BNA OSHC 1692, 1703 (No. 79-1998, 1986) (holding that though OSHA's suggested means of abatement appeared as though it may "substantially interfere" with an employer's operations, because the employer "made no attempt" to rebut OSHA's suggested feasible means of abatement, OSHA successfully met its burden of showing that a feasible means of abatement exists).

[58] Waldon Health Care Ctr., 16 BNA OSHC 1052, 1066 (No. 89-3097, 1993) (observing that proposed methods of abatement that are cost prohibitive are not feasible because "under the general duty clause, an employer is not required to adopt measures that would threaten its economic viability"); Inland Steel Co., No. 79-3286 at *10-11 (OSHRC Apr. 7, 1982).

[59] Steel Constructors, Inc., 8 BNA OSHC 2146, 2151 (No. 78-3839, 1980) (vacating a General Duty Clause citation because OSHA's proposed means of abatement "amounted to a greater hazard than that method followed by [employer] ...").

[60] Cargill, Inc., 10 BNA OSHC 1398, 1401 (No. 78-5707, 1982).

[61] Id. at 1407.

[62] Id. at 1408.

[63] PSP Monotech Indus., 22 BNA OSHC 1303, 1308 (No. 06-1201, 2008); Tampa Shipyards, Inc., 15 BNA OSHC 1533, 1537-38 (Nos. 86-360 and 86-469, 1992).

[64] W.G. Yates and Sons Constr. Co., Inc. v. OSHRC, 459 F.3d 604, 607 (5th Cir. 2006) (observing that the OSH Act sought prevention of only preventable and foreseeable hazards, and thus employers can only be liable if they knew, or with the exercise of reasonable diligence should have known, of the presence of a violation); Brennan v. OSHRC, 511 F.2d 1139, 1144-45 (9th Cir. 1975) (holding that the requirement of employer knowledge applies to all degrees of violations of the OSH Act, whether a violation is "non-serious," "serious," or "willful.").

[65] See Sofco Erectors, Inc., 12 BNA OSHC 1867, 1868 (No. 85-776, 1986) (holding that an employer's argument that he lacked knowledge of a condition failed in part because "[i]f [employer] lacked knowledge, it was directed to the requirements of the law but not to the physical conditions which constitute a violation.").

[66] See id.

[67] See, e.g., J-Lenco, Inc., 19 BNA OSHC 1992, 1994 (No. 01-0712, 2002) (holding that OSHA met its burden of showing actual employer knowledge through employer admissions).

[68] Tampa Shipyards, 15 BNA OSHC at 1548.

[69] PSP Monotech Indus., 22 BNA OSHC 1303, 1308 (No. 06-1201, 2008).

[70] R.P. Carbone Constr. Co., 18 BNA OSHC 1056, 1064 (No. 96-1302, 1998).

[71] See id.

[72] Id. at 1058.

[73] Id. at 1064.

[74] Tampa Shipyards, 15 BNA OSHC at 1548; ComTran Grp., Inc. v. DOL, 722 F.3d 1304, 1317 (11th Cir. 2013); New York State Elec. and Gas Corp. v. Sec'y of Labor, 88 F.3d 98, 105 (2nd Cir. 1996).

[75] Tampa Shipyards, 15 BNA OSHC at 1548.

[76] W.G. Yakes and Sons Constr. Co., Inc. v. OSHRC, 459 F.3d 604, 608-09 (5th Cir. 2006).

[77] Id.

[78] 29 U.S.C. §654(a)(2).

79 29 U.S.C. §652(8).

80 For example, the standards found in Title 29, Part 1918 of the Code of Federal Regulations (CFR) apply only to the longshoring industry, and the standards found in Title 29, Part 1926 of the CFR apply only to the construction industry.

81 General standards can be found in Title 29, Part 1910 of the CFR.

82 Gary Concrete Prods., Inc., 15 BNA OSHC 1051, 1051 (No. 86-1087, 1991).

83 See id.

84 Id.

85 Pico Indus., 21 BNA OSH 1228, 1229 (No. 04-1419, 2005).

86 29 C.F.R. §1910.5(c)(1) ("if a particular standard is specifically applicable to a condition, practice, means, method, operation, or process, it shall prevail over any different general standard which might otherwise be applicable to the same condition, practice, means, method, operation, or process.").

87 See, e.g., see Anthony Crane Rental v. Reich, 70 F.3d 1290, 1302 (D.C. Cir. 1995) (observing that "the general industry standards apply unless they are preempted by specific industry standards.").

88 See 29 U.S.C. §654(a)(2).

89 See id.

90 See Indus. Steel Erectors, Inc., 1 BNA OSHC 1497, 1498 (No. 703, 1974).

91 See id; see also 29 C.F.R. §1905.5.

92 29 U.S.C. §§655, 665.

93 29 U.S.C. §655(b)(6)(A).

94 Id.

95 Id.

96 Id.

97 Id.

98 Id.

99 Id.

100 29 U.S.C. §655(d).

101 See id.

102 Id.

103 Id.

104 Id.

105 See id.

106 Id.

107 Id.

108 29 U.S.C. §655(b)(6)(C).

109 Id.

110 29 U.S.C. §665.

111 Id.

112 Id.

113 See 29 U.S.C. §657(c)(2).

114 See 29 U.S.C. §657(c)(2); see also 29 U.S.C. §673(a).

115 See 29 C.F.R. pt. 1904.

116 See 29 C.F.R. §1904.1-2.

117 29 C.F.R. §1904.1.

118 29 C.F.R. §1904.2(a)(1).

119 29 C.F.R. §1904 App'x A.

[120] 29 C.F.R. §1904.46.

[121] 29 C.F.R. §1904.2(a)(2).

[122] 29 C.F.R. §1904.41, 1904.42.

[123] See 29 C.F.R. §1904.4(a).

[124] Id.

[125] 29 C.F.R. §1904.5(a).

[126] Id. However, the regulations do contain exceptions to this presumption. See 29 C.F.R. §1904.5(b)(2). These exceptions include, but are not limited to, injuries or illnesses resulting from personal grooming or the employee doing personal tasks at a workplace outside of assigned working hours. Id.

[127] 29 C.F.R. §1904.6(a)(1), (2).

[128] 29 C.F.R. §1904.6(a)(2).

[129] 29 C.F.R. §1904.7(a).

[130] The reporting requirements applicable to inspections are discussed in the "Inspections" section of this report.

[131] See 29 C.F.R. §1904.39. The reporting requirements for serious injuries were expanded via a regulation that became effective January 1, 2015. See Occupational Injury and Illness Recording and Reporting Requirements—NAICS Update and Reporting Revisions, 79 Fed. Reg. 56187 (Sept. 18, 2014).

[132] 29 C.F.R. §1904.39(a). This reporting requirement does not extend to deaths or injuries caused by motor vehicle accidents on public streets or highways that are not in construction zones. 29 C.F.R. §1904.39(b)(3). Additionally, this reporting requirement does not apply to deaths or injuries involving commercial plane, train, subway, or bus accidents. 29 C.F.R. §1904.39(b)(4).

[133] 29 C.F.R. §1904.39(a)(1).

[134] 29 C.F.R. §1904.39(a)(2).

[135] 29 C.F.R. §1904.39(a)(3).

[136] See 29 C.F.R. §1904.41.

[137] See 29 C.F.R. §1904.41(a), (b)(1).

[138] Id.

[139] 29 C.F.R. §1904.41(b)(3).

[140] See Improve Tracking of Workplace Injuries and Illnesses, 78 Fed. Reg. 67254 (proposed Nov. 8, 2013). After multiple extensions, the comment period for this proposed rule closed in October 2014. Supplemental Notice of Proposed Rulemaking, 79 Fed. Reg. 47605 (Aug, 14, 2014).

[141] Id.

[142] Id.

[143] Id.

[144] Id.

[145] Id.

[146] Id.

[147] 29 C.F.R. §1903.2(a)(1).

[148] 29 C.F.R. §1904.32.

[149] 29 C.F.R. §1903.16(a).

[150] 29 C.F.R. §1903.14a(c)(1).

[151] 29 U.S.C. §657(c)(1).

[152] 29 C.F.R. §1903.2(a)(1).

[153] OSHA, JOB SAFETY AND HEALTH: IT'S THE LAW (2012), available at https://www.osha.gov/Publications/poster.html.
[154] 29 C.F.R. §1903.2(a)(1).
[155] Id.
[156] 29 C.F.R. §1904.32.
[157] 29 C.F.R. §1904.32(b)(5).
[158] Id.
[159] 29 C.F.R. §1903.16(a). This posting requirement does not apply to notices of de minimis OSH Act violations. Id.
[160] Id.
[161] Id.
[162] 29 C.F.R. §1903.16(b).
[163] Id.
[164] 29 C.F.R. §1903.16(c). The notice can include any specific actions the employer has taken to abate the violation. Id.
[165] See 29 C.F.R. §1903.19.
[166] 29 C.F.R. §1903.14a(a).
[167] 29 C.F.R. §1903.14a(c)(1).
[168] Id.
[169] See 29 U.S.C. §657(a).
[170] Id.
[171] See id.
[172] Id.
[173] Sec'y of Labor v. Barlow's, Inc., 436 U.S. 307 (1978).
[174] See Cody Zeigler, Inc., Nos. 00-0040, 00-0222 at *3 (OSHRC Nov. 12, 2000).
[175] See id.
[176] Trinity Indus., Inc., Nos. 88-1545, 188-1547 at *16 (OSHRC Aug. 10, 1990); Gem Indus., Inc. 17 BNA OSHC 1184, 1186 (No. 93-1122, 1995).
[177] VSH Rest. Corp., 17 BNA OSHC 1336, 1337 (No. 93-1357, 1995).
[178] Id.
[179] See Envtl. Utils. Corp., 5 BNA OSHC 1195,1196 (No. 5324, 1977).
[180] Sec'y of Labor v. Barlow's, Inc., 436 U.S. 307 (1978).
[181] Id. at 309.
[182] Id. at 312.
[183] Id.
[184] Id. at 309.
[185] Id. at 325. The Court observed that, in the context of an OSHA inspection, the probable cause necessary to support a warrant can take one of two forms: (1) it can be based on "specific evidence of an existing violation" of the OSH Act; or (2) it can be based on OSHA showing that reasonable standards for inspecting the workplace are satisfied. Id. at 320.
[186] See Shneckloth v. Bustamonte, 412 U.S. 218, 219 (1973) ("it is ... well settled that one of the specifically established exceptions to the requirements of both a warrant and probable cause is a search that is conducted pursuant to consent.").
[187] See OSHA, OSHA'S FIELD OPERATIONS MANUAL 2-8 (2011), available at https://www.osha.gov/ OshDoc/Directive_pdf/CPL_02-00-150.pdf.
[188] See id.
[189] 29 U.S.C. §657(f)(1).
[190] Id.

[191] Id.
[192] Id.
[193] See id.
[194] 29 C.F.R. §1903.12(a).
[195] Id.
[196] See OSHA, OSHA'S FIELD OPERATIONS MANUAL 9-2, 9-3 (2011), available at https://www.osha.gov/ OshDoc/Directive_pdf/CPL_02-00-150.pdf.
[197] Id. at 9-3.
[198] Id.
[199] 29 C.F.R. §1904.39(a).
[200] OSHA, OSHA'S FIELD OPERATIONS MANUAL 11-7 (2011), available at https://www.osha.gov/ OshDoc/Directive_pdf/CPL_02-00-150.pdf.
[201] See OSHA, OSHA'S FIELD OPERATIONS MANUAL (2011) at Chapter 2 part VI, available at https://www.osha.gov/ OshDoc/Directive_pdf/CPL_02-00-150.pdf.
[202] See, e.g., OSHA, INSPECTION SCHEDULING FOR CONSTRUCTION (2013).
[203] See OSHA, OSHA'S FIELD OPERATIONS MANUAL 2-10 (2011), available at https://www.osha.gov/OshDoc/ Directive_pdf/CPL_02-00-150.pdf.
[204] 29 U.S.C. §666(f).
[205] 29 C.F.R. §1903.6(a).
[206] 29 C.F.R. §1903.7(a).
[207] OSHA, OSHA'S FIELD OPERATIONS MANUAL 3-6 (2011), available at https://www.osha.gov/OshDoc/Directive_pdf/CPL_02-00-150.pdf.
208 29 C.F.R. §1903.7(a).
209 29 C.F.R. §1904.40(a).
210 29 C.F.R. §1903.7(a).
211 29 U.S.C. §657(e).
212 29 C.F.R. §1903.8(a).
213 29 C.F.R. §1903.8(c).
214 Id.
215 29 C.F.R. §1903.8(d).
216 29 C.F.R. §1903.7(b); OSHA, OSHA'S FIELD OPERATIONS MANUAL 3-22 (2011), available at https://www.osha.gov/OshDoc/Directive_pdf/CPL_02-00-150.pdf.
217 29 C.F.R. §1903.10.
218 29 C.F.R. §1903.13.
219 OSHA, OSHA'S FIELD OPERATIONS MANUAL 11-3 (2011), available at https://www.osha.gov/OshDoc/Directive_pdf/CPL_02-00-150.pdf.
220 29 U.S.C. §662.
221 29 C.F.R. §1903.7.
222 29 U.S.C. §664.
223 Id.
224 Id.
225 OSHA, OSHA'S FIELD OPERATIONS MANUAL 3-28 (2011), available at https://www.osha.gov/OshDoc/Directive_pdf/CPL_02-00-150.pdf.
226 Id.
227 Id.
228 Id.
229 29 C.F.R. §1903.14(a).
230 29 U.S.C. §658(a).

[231] 29 U.S.C. §658(a), (c). OSHRC has interpreted the requirement that any citation be issued within six months of the occurrence of a violation to mean that citations must be issued within six months from when the Secretary discovers, or reasonably should have discovered, that the violation exists. See Turner Constr. Co, 19 BNA OSHC 1926, 1929 (No. 01-0814, 2002); see also Austin Indus. Specialty Servs., No. 11-2555 at *15 (OSHRC June 4, 2013).

[232] 29 U.S.C. §658(a).

[233] 29 C.F.R. §1903.15(a).

[234] Id.

[235] Id.

[236] 29 U.S.C. §666(k).

[237] See Wal-Mart Stores v. Sec'y of Labor, 406 F.3d 731, 735 (D.C. Cir. 2005); Phelps Dodge Corp. v. OSHRC, 725 F.2d 1237, 1240 (9th Cir. 1984) (holding erroneous the defendant's contention that serious violations require the potential to cause physical harm); Alabama Salvage Auction Co. at *6 (OSHRC Mar. 3, 2014) (observing that to establish a serious violation "the Secretary need not show that there was a substantial probability that an accident would actually occur; he need only show that if an accident occurred, serious physical harm could result.").

[238] Wal-Mart Stores, 406 F.3d at 735.

[239] 29 U.S.C. §666(k).

[240] 29 U.S.C. §666(b).

[241] 29 U.S.C. §666(b), (c).

[242] 29 U.S.C. §666(c).

[243] See, e.g., S and F Concrete Contractors, 20 BNA OSHC 1943, 1942 (No. 03-1816, 2004); Lourdes Hosp., 20 BNA OSHC 1789, 1795 (No. 03-0641, 2004).

[244] 29 U.S.C. §666(c).

[245] 29 U.S.C. §658(a).

[246] OSHA, OSHA'S FIELD OPERATIONS MANUAL 4-36 (2011), available at https://www.osha.gov/OshDoc/Directive_pdf/CPL_02-00-150.pdf.

[247] 29 U.S.C. §658(a); 29 C.F.R. §1903.14(a).

[248] Gen. Carbon Co. v. OSHRC, 860 F.2d 479, 487 (D.C. Cir. 1988).

[249] See 29 C.F.R. §1903.14(a).

[250] 29 C.F.R. §1903.15(c).

[251] 29 U.S.C. §666(a).

[252] See, e.g., Conie Constr., Inc. v. OSHRC, 73 F.3d 382, 384 (D.C. Cir. 1995); Daniel v. OSHRC, 295 F.3d 1232, 1239 (11th Cir. 2002); Trinity Indus., Inc. v. OSHRC, 107 Fed. App'x 387, 393 (5th Cir. 2004); 23 BNA OSHC 1596, 1620 (No. 94-3393, 2011).

[253] See, e.g., Wynnewoood Ref. Co. v. Sec'y of Labor, 340 Fed. App'x 462, 466 (10th Cir. 2009) (upholding OSHRC's determination that an employer willfully violated the act when it knowingly failed to comply with a standard because of economic infeasibility); Donovan v. Williams Enters., Inc., 744 F.2d 170, 180 (D.C. Cir. 1984) (upholding OSHRC's determination that an employer willfully violated the act when OSHA repeatedly warned the employer about his failure to comply with safety standards and the employer ignored the warnings).

[254] John Carlo, Inc., 21 BNA OSHC 1670, 1673 (No. 04-1405, 2006).

[255] Chao v. Barbosa Grp., Inc., 296 Fed. App'x 211, 213 (2nd Cir. 2008); (Gen. Motors Corp., 14 BNA OSHC 2064, 2068 (Nos. 82-630, 84-781, 84-816, 1991).

[256] St. Joe Minerals Corp. v. OSHRC, 647 F.2d 840, 847-48 (8th Cir. 1981); Mobil Oil Corp., 10 BNA OSHC 1606, 1615 (No. 80-6678, 1982) (observing that OSHA's burden of establishing willful conduct in the context of a General Duty Clause violation is "notably more difficult.").

[257] 29 U.S.C. §666(a).

[258] 29 U.S.C. §666(e).

[259] Id.

[260] 29 C.F.R. §1903.18(a).

[261] A citation generally becomes a final order through settlement, litigation, or, if uncontested by the employer within 15 working days of receiving a notice of proposed penalty pursuant to a citation, through operation of law pursuant to 29 U.S.C. §659(a).

[262] M and J Painting Co., Inc., 24 BNA OSHC 1450, 1468 (No. 11-2694, 2012); Hercules, Inc., 20 BNA OSHC 2097, 2098 (No. 94-2790, 2005).

[263] Hercules, Inc., 20 BNA OSHC 2097, 2099-2100 (No. 94-2790, 2005) (finding that because a citation for failure to properly record injuries and illnesses did not provide sufficient information to allow the employer to bring his records into compliance, the citation lacked sufficient "particularity" for a subsequent failure-to-abate violation).

[264] Franklin Lumber Co., Inc, 2 BNA OSHC 1077, 1077 (No. 900, 1974); M and J Painting, Co., 24 BNA OSHC 1450, 1467 (No. 11-2694, 2012).

[265] See 29 U.S.C. §666(d).

[266] See Midwest Masonry, Inc., 19 BNA OSHC 1540, 1541 (No. 00-322, 2001).

[267] See id; see also Potlach Corp., No. 16183 at *3 (OSHRC Jan. 22, 1979).

[268] Bunge Corp. v. Sec'y of Labor, 638 F.2d 831, 837 (5th Cir. 1981); Potlach Corp., No. 16183 at *3 (OSHRC Jan. 22, 1979).

[269] See Bunge Corp., 638 F.2d at 837; see also Potlach Corp., No. 16183 at *3 (OSHRC Jan. 22, 1979).

[270] See Bunge Corp., 638 F.2d at 837; see also J.L. Foti Constr. Co. v. OSHRC, 687 F.2d 853, 857 (6th Cir. 1982); Potlach Corp., No. 16183 at *3 (OSHRC Jan. 22, 1979).

[271] See Potlach Corp., No. 16183 at *4 (OSHRC Jan. 22, 1979).

[272] 29 U.S.C. §666(a).

[273] 29 U.S.C. §659(a).

[274] Id.

[275] 29 U.S.C. §659(c).

[276] 29 U.S.C. §659(a).

[277] 29 U.S.C. §659(c).

[278] Id.

[279] See 29 U.S.C. §661(j).

[280] 29 C.F.R. §2200.90(a).

[281] 29 C.F.R. §2200.91(b).

[282] 29 C.F.R. §2200.91(a).

[283] Id.

[284] 29 C.F.R. §2200.90(d).

[285] See 29 C.F.R. §2200.94.

[286] 29 U.S.C. §660(a).

[287] 29 U.S.C. §660(b).

[288] 29 U.S.C. §660(a), (b).

[289] See 29 U.S.C. §660(a).

[290] 29 U.S.C. §660(a).

In: The Occupational Safety and Health Act
Editor: Christy Knight
ISBN: 978-1-63482-637-2

Chapter 2

OSHA STATE PLANS: IN BRIEF, WITH EXAMPLES FROM CALIFORNIA AND ARIZONA*

Scott D. Szymendera

SUMMARY

The Occupational Safety and Health Act of 1970 (OSH Act) authorizes states to establish their own occupational safety and health plans and preempt standards established and enforced by the Occupational Safety and Health Administration (OSHA). OSHA must approve state plans if they are "at least as effective" as OSHA's standards and enforcement. Currently, 21states and Puerto Rico have state plans that cover all employers, whereas four states and the U.S. Virgin Islands have state plans that cover only state and local government employers, who are not covered by the OSH Act. OSHA estimates that 40% of all American workers are covered by state occupational safety and health plans.

California's state plan includes specific standards designed to protect workers against heat illness despite OSHA not having a federal heat illness standard, exemplifying a state plan that goes beyond OSHA's standards. In contrast, recent legislation enacted in Arizona creates a

* This is an edited, reformatted and augmented version of a Congressional Research Service publication R43969, prepared for Members and Committees of Congress dated April 3, 2015.

residential construction fall protection standard that OSHA has determined is not as effective as the federal standard and has been formally rejected by OSHA, putting the approval of Arizona's state occupational safety and health plan into potential jeopardy.

STATE OCCUPATIONAL SAFETY AND HEALTH PLANS

Under the provisions of the Occupational Safety and Health Act of 1970 (OSH Act),[1] the federal government, through the Occupational Safety and Health Administration (OSHA), has the primary responsibility for establishing and enforcing workplace safety standards.[2] However, Section 18 of the OSH Act permits any state or territory to preempt federal jurisdiction over occupational safety and health by establishing an approved state occupational safety and health plan.[3]

As shown in *Figure 1* and *Table 1*, 21 states and Puerto Rico currently have state plans that cover private- and public-sector employers whereas four states and the U.S. Virgin Islands have plans that cover only state and local government employers, who are not covered by OSHA. In addition, Maine is currently in the process of establishing a state plan for state and local government employers, and New Jersey, which has a state plan for public-sector workers only, is in the process of extending its state plan to cover all employers. OSHA estimates that 40% of all American workers are covered by a state occupational safety and health plan.[4]

State Plan Requirements

In general, for a state occupational safety and health plan to be approved, it must be "at least as effective" as the standards established and enforced by OSHA at providing for workplace safety. Specifically, Section 18(c) of the OSH Act requires OSHA to approve a state plan if the plan

- designates a state agency or agencies that are responsible for administering the plan throughout the state;

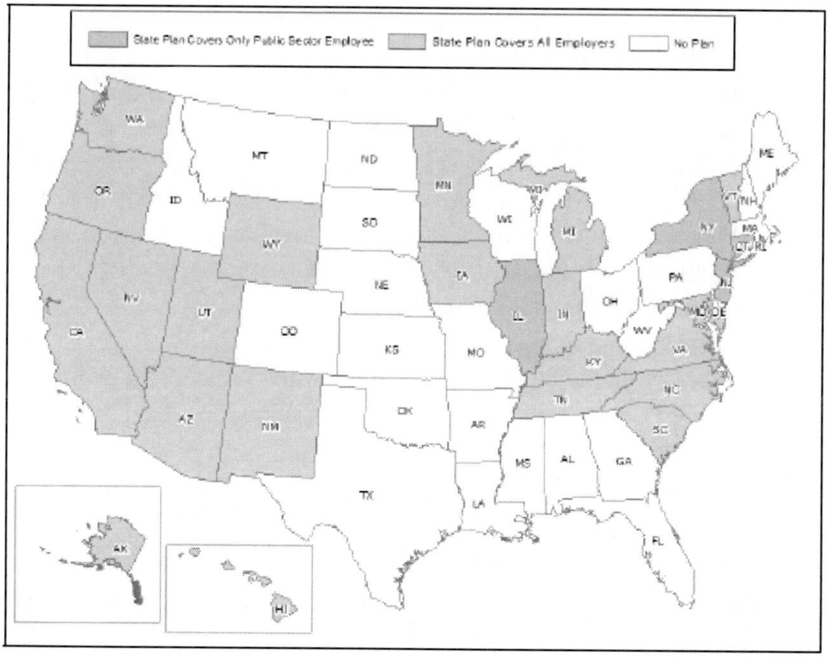

Source: Occupational Safety and Health Administration (OSHA) website at https://www.osha.gov/dcsp/osp/ indes/html.

Note: Does not include territories.

Figure 1. OSHA-Approved State Occupational Safety and Health Plans.

Table 1. OSHA-Approved State Occupational Safety and Health Plans

Covers Private- and Public-Sector Workers	Covers Only Public-Sector Workers	
Alaska	New Mexico	Connecticut
Arizona	North Carolina	Illinois
California	Oregon	New Jersey
Hawaii	Puerto Rico	New York
Indiana	South Carolina	U.S. Virgin Islands
Iowa	Tennessee	
Kentucky	Utah	
Maryland	Vermont	
Michigan	Virginia	
Minnesota	Washington	
Nevada	Wyoming	

Source: Occupational Safety and Health Administration (OSHA) website at https://www.osha.gov/dcsp/osp/ index.html.

- provides for the development and enforcement of safety and health standards relating to one or more safety or health issues, which standards (and the enforcement of such standards) are or will be at least as effective in providing safe and healthful employment and places of employment as the standards promulgated under Section 6 [of the OSH Act] relating to the same issues, and which standards, when applicable to products that are distributed or used in interstate commerce, are required by compelling local conditions and do not unduly burden interstate commerce;
- provides for a right of entry and inspection of all workplaces subject to the act that is at least as effective as that provided in Section 8 [of the OSH Act], and includes a prohibition on advance notice of inspections;
- contains satisfactory assurances that such agency or agencies have or will have the legal authority and qualified personnel necessary for the enforcement of such standards;
- gives satisfactory assurances that such state will devote adequate funds to the administration and enforcement of such standards;
- contains satisfactory assurances that such state will, to the extent permitted by its law, establish and maintain an effective and comprehensive occupational safety and health program applicable to all employees of public agencies of the state and its political subdivisions, which program is as effective as the standards contained in an approved plan;
- requires employers in the state to make reports to the Secretary [of Labor] in the same manner and to the same extent as if the plan were not in effect; and
- provides that the state agency will make such reports to the Secretary in such form and contain such information, as the Secretary shall from time to time require.[5]

State Plan Approval Process

Any state or territory may submit a state plan to OSHA for approval. The first step in the approval process is for a state to submit a *Developmental Plan* to OSHA. Required elements of a developmental plan include

- state legislation,
- regulations and procedures for establishing occupational safety and health standards,
- enforcement systems,
- appeals processes, and
- sufficient staffing to implement the state plan.

In addition, a developmental plan must demonstrate that within three years, the state will have the necessary infrastructure to be effective.

The state plan is eligible for *Certification* once a state has submitted a developmental plan. OSHA provides certification once a state demonstrates that it has the necessary infrastructure in place as outlined in its developmental plan. The OSHA Certification process does not render any judgment on the effectiveness of the state plan and does not result in the preemption of federal jurisdiction.

When OSHA determines that a state is capable of meeting the statutory state plan requirements, including the requirement that the plan is "at least as effective" as the federal system, OSHA and the state may enter into an *Operational Status Agreement* that outlines which employers will be under state jurisdiction and which will remain under federal jurisdiction.

A state plan may receive *Final Approval* at least one year after certification. Final approval is the formal decision of OSHA that the plan is meeting its statutory requirements and that the plan is as least as effective as the federal system. After final approval is granted, all employers, except those under federal jurisdiction as a matter of law, are under state, rather than federal, jurisdiction.[6]

A state dissatisfied with OSHA's decision not to grant approval to its state plan may request a hearing before an administrative law judge of the Occupational Safety and Health Review Commission (OSHRC).

A plan may operate indefinitely under an operational status agreement without ever receiving final approval from OSHA, which may be modified upon agreement between the state and OSHA. For example, the largest state plan, administered by the California Division of Occupational Safety and Health (Cal/OSHA) received certification in 1977 and has been operating under an operational status agreement since that time without having received final approval from OSHA. Currently, seven states and territories are operating state plans under operational status agreements rather than final approvals.[7]

Monitoring of State Plans

OSHA monitors approved state plans to ensure that they remain at least as effective as the federal system in providing for occupational safety and health through the Federal Annual Monitoring Evaluation (FAME) process. Monitoring state plans includes (1) reviewing state laws and standards to ensure effectiveness; (2) reviewing the operation and administration of state plans; and (3) assessing state progress toward meeting performance goals. Each year, OSHA issues a FAME report for each state plan.[8] If OSHA finds deficiencies in a state's plan, it can require changes to the plan and ultimately reconsider its decision to approve the state plan such that OSHA reasserts partial or full jurisdiction over occupational safety and health in the state.

State Plan Funding

Section 23(g) of the OSH Act authorizes the Department of Labor (DOL) to make matching grants to states to cover up to 50% of the cost of the administration of approved state occupational safety and health plans.[9] A state may overmatch the federal grant amount with state funds. The total amount available for state plan grants comes from the discretionary appropriation to the DOL for OSHA. The FY2015 appropriation for OSHA state plan grants was $1.850 million and DOL requested $104.377 million for state plan grants for FY2016.[10]

EXAMPLES OF STATE OCCUPATIONAL SAFETY AND HEALTH PLANS

Detailed information on each of the approved state occupational safety and health plans is available on the OSHA website.[11] This section will focus on two state plans that are involved with current controversial occupational safety and health issues. The California state plan exceeds OSHA's standards and includes specific requirements that employers must fulfill when their employees are working outdoors during periods of high temperatures. The Arizona state plan includes a residential construction fall protection standard that has recently been deemed insufficient by OSHA and may result in OSHA reconsidering its approval of the state plan.

California

The California state occupational safety and health plan is administered by Cal/OSHA. The California plan is unique in that it includes a specific occupational safety and health standard that applies in cases of employees working outdoors during periods of high temperatures with the goal of reducing heat-related illnesses. Unlike the California state plan, OSHA does not have a specific standard addressing the hazards of working outdoors in high temperatures. Thus, the California state plan exemplifies going beyond the statutory requirement of providing a state plan that is at least as effective as the federal system and implementing standards to address hazards not specifically covered by OSHA.

OSHA and Heat Illness

There is no federal occupational safety and health standard that specifically addresses the hazards of working outdoors in high temperatures. Rather, OSHA relies on the following existing section of the OSH Act and federal standards to attempt to reduce heat-related illnesses and injuries:

- Section 5(a)(1) of the OSH Act, commonly referred to as the General Duty Clause, requires that each employer
- shall furnish to each of his employees employment and a place of employment which are free from recognized hazards that are causing or are likely to cause death or serious physical harm to his employees;[12]
- the personal protective equipment (PPE) standards require each employer to assess potential hazards to employees and provide employees with appropriate PPE to protect against these hazards;[13]
- the sanitation standards require employers to provide employees with access to potable water;[14] and
- the safety training and education standard for construction requires construction employers to "instruct each employee in the recognition and avoidance of unsafe conditions" in the workplace.[15]

In addition, OSHA has developed a smartphone application that allows employers and workers to determine the current heat index and risk level for heat illness and provides recommended protection measures based on the current risk level. OSHA has also launched a heat illness prevention safety campaign that includes the use of materials, such as the poster shown in *Figure 2*, developed by Cal/OSHA as part of California's state occupational safety and health plan.

Cal/OSHA's Heat Illness Prevention Standard

In 2005, Cal/OSHA established a heat illness prevention standard as part of California's state occupational safety and health plan.[16] The Cal/OSHA standard requires all employers with outdoor places of employment to

- provide training in the risks and prevention of heat illness to all employees and supervisors before engaging in work that may expose employees to high temperatures; and
- in temperatures of at least 85 degrees Fahrenheit, provide access to shade or other celling systems, such as misters, and permit employees to take breaks of at least five minutes whenever they feel it necessary to protect against heat illness.

Source: OSHA website at https://www.osha.gov/SLTC/heatillness/edresources.html.

Figure 2. OSHA Heat Illness Poster Developed by Cal/OSHA.

The Cal/OSHA standard has additional rules regarding periods of high heat (at least 95 degrees Fahrenheit) that apply to the following industries:

- agriculture,
- construction, landscaping,
- oil and gas extraction, and
- transportation of agricultural production, construction materials, or other heavy items unless the employee is in an air-conditioned vehicle and is not involved in the loading or unloading of the vehicle.

At temperatures of at least 95 degrees Fahrenheit, covered employers must maintain communication with all employees; observe employees for signs of heat illness, especially, new employees not yet acclimated to high heat conditions; and remind employees to drink water.[17]

Arizona

The Arizona state occupational safety and health plan received final approval from OSHA in 1985 and is administered by the Arizona Occupational Safety and Health Division (ADOSH) of the Industrial Commission of Arizona (ICA). In February 2015, OSHA announced that it was rejecting the residential construction fall protection standard in the Arizona state plan and that, pending additional action by the state of Arizona, may reconsider its final approval of the plan.

OSHA's Residential Construction Fall Protection Standard

OSHA's residential construction fall protection standard requires that each employee who works in residential construction activities at a height of at least six feet above the ground or a lower level be protected from falls by a guardrail, safety net, or personal fall arrest system (known as conventional fall protection).[18] This standard was initially issued on August 1994, but concerns over its feasibility led OSHA to issue interim compliance procedures in December 1995[19]that permitted employers to use non-conventional fall protection systems, such as slide guards, rather than the safety systems required by the standard. On December 16, 2010, OSHA cancelled its interim guidance and announced that the residential construction fall protection standard would go into effect in June 2011.[20] This effective date was later postponed by OSHA to September 22, 2011.[21]

Arizona's Residential Construction Fall Protection Legislation

On June 16, 2011, ADOSH adopted the federal residential fall protection standard. However, the next day the ICA stayed enforcement of the standard. The stay was lifted on November 20, 2011,effective January 1, 2012. On March 27, 2012, SB 1441 was enacted into law in Arizona.[22] The legislation requires conventional fall protection only when an employee is working at a height of at least 15 feet or when the slope of the roof is steeper than 7:12 (measured as the ratio of rise to run). The legislation also created an exception in cases in which conventional fall protection is not feasible or creates a greater hazard.[23] ADOSH adopted the provisions of SB 1441 as a standard as part of its state occupational safety and health plan.[24]

On April 22, 2014, Arizona enacted SB 1307, which made minor changes to the residential construction fall protection standard while maintaining the minimum height of 15 feet for the use of conventional fall protection. In addition, Section 7 of SB 1307 includes a provision that repeals the provisions in the Arizona Revised Statutes created by SB 1441 that contain the residential construction fall protection standard if OSHA publishes a notice rejecting the state statute.

OSHA's Response to Arizona's Legislation

On March 19, 2014, OSHA sent the Arizona ICA a show-cause letter asking the ICA to demonstrate why OSHA should not begin proceedings to reject the state's residential construction fall protection statute and reconsider final approval of the Arizona state plan on the grounds that the standard is not at least as effective as the federal standard. Arizona's response to this letter focused on the changes to the initial legislation made by the enactment of SB 1307.

On February 6, 2015, OSHA published a notice in the *Federal Register* that it was formally rejecting the Arizona residential construction fall protection standard and excluding this standard from Arizona's state plan.[25] Because of the provision in SB 1307 repealing the Arizona residential construction fall protection legislation upon notice of its rejection by OSHA, OSHA stated that it expects Arizona to now incorporate the federal standard into its state plan and will defer any decision on reconsidering final approval of the Arizona state plan pending the state's action.

If OSHA determines that Arizona is not enforcing a residential construction fall protection standard that is at least as effective as the federal standard, then OSHA could begin the process of withdrawing final approval of all or part of the Arizona state occupational safety and health plan.[26] If final approval of the entire plan were to be withdrawn by OSHA, OSHA would then take jurisdiction over all employers in Arizona unless an operational status agreement between Arizona and OSHA was signed. If final approval of only part of the plan were to be withdrawn, then OSHA would take jurisdiction over employers in the withdrawn areas of the plan and the rest of employers in the state would remain under ADOSH jurisdiction.

End Notes

[1] 29 U.S.C. §§ 651et seq. For additional information on the Occupational Safety and Health Act of 1970 (OSH Act), see CRS Report R43768, The OSH Act: A Legal Overview, by Rodney M. Perry.

[2] Under Section 4(b)(1) of the OSH Act, the Occupational Safety and Health Administration (OSHA) does not have jurisdiction over areas in which another federal agency has exercised its authority to establish and enforce workplace health and safety standards.

[3] 29 U.S.C. § 667.

[4] U.S. Department of Labor (DOL), FY 2016 Congressional Budget Justification: Occupational Safety and Health Administration, February 2015, p. 59, http://www.dol.gov/dol/budget/2016/PDF/CBJ-2016-V2-12.pdf.

[5] 29 U.S.C. § 667(c).

[6] For example, under Section 19 of the OSH Act [29 U.S.C. § 668], the occupational safety and health programs required to be maintained by federal agencies remain under the jurisdiction of OSHA.

[7] State plans in California, Hawaii, Michigan, New Mexico, Puerto Rico, Vermont, and Washington are currently operating under Operational Status Agreements rather than Final Approvals.

[8] Federal Annual Monitoring Evaluation (FAME) reports are available on the OSHA website at https://www.osha.gov/ dcsp/osp/efame/index.html.

[9] 29 U.S.C. § 672(g).

[10] DOL, FY 2016 Congressional Budget Justification: Occupational Safety and Health Administration, February 2015, p. 68, http://www.dol.gov/dol/budget/2016/PDF/CBJ-2016-V2-12.pdf.

[11] For detailed information on approved state plans, see https://www.osha.gov/dcsp/osp/approved_state_plans.html.

[12] 29 U.S.C. § 654(a)(1).

[13] 29 C.F.R. §§ 1910.132, 1915.152, 1917.95, and 1926.28.

[14] 29 C.F.R. §§ 1910.141, 1915.88, 1917.127, 1918.95, 1926.51, and 1928.110.

[15] 29 C.F.R. § 1926.21.

[16] The Cal/OSHA heath illness prevention standard is codified at Cal. Code Regs., Tit. 8, § 3395 and is available at https://www.dir.ca.gov/Title8/3395.html.

[17] Amendments to the Cal/OSHA heat illness prevention standard that would provide additional protections to workers have been proposed by the California Occupational Safety and Health Board. If these amendments are approved by the California Office of Administrative Law, they will go into effect on May 1, 2015. Information on these amendments is available

at https://www.dir.ca.gov/dosh/documents/Heat-Illness-Prevention-Regulation-Amendments.pdf.

[18] 29 C.F.R. § 1926.501(b)(13).

[19] OSHA Instruction STD 3.1. On June 18, 1999, OSHA issued Directive # STD 3-00-001, a plain language revision of Instruction STD 3.1.

[20] OSHA Directive # STD 03-11-002.

[21] This postponement was announced in OSHA's letter of June 8, 2011, to the National Association of Home Builders. This letter is available at https://www.osha.gov/doc/letter_nahb.html.

[22] This legislation was codified at A.R.S. 23-492.

[23] One argument against the use of conventional fall protection systems in residential construction is that the wood commonly used in residential construction does not provide a safe anchor point for personal fall arrest systems. See, for example, the comments of Spencer Kamps of the Home Builders Association of Central Arizona in Carina Dominguez, "Feds Reject Ariz. Construction Worker Safety Plan," azcentral, March 5, 2015, at http://www.azcentral.com/story/money/real-estate/2015/03/05/feds-reject-ariz-construction-worker-safety-plan/24404521/.

[24] The Arizona residential construction fall protection standard is codified at Ariz. Admin. Code R20-5-601.01.

[25] DOL, "Arizona State Plan for Occupational Safety and Health," 80 Federal Register 6652, February 6, 2015. This final notice followed a proposed notice and solicitation of public comments published on August 21, 2014 (79 Federal Register 49465).

[26] Procedures for the withdrawal of all or a portion of a state occupational safety and health plan are provided in Part 1955 of Title 29 of the Code of Federal Regulations (C.F.R.).

In: The Occupational Safety and Health Act
Editor: Christy Knight
ISBN: 978-1-63482-637-2

Chapter 3

ALL ABOUT OSHA*

Occupational Safety and Health Administration

This publication provides a general overview of basic topics related to OSHA and how it operates. Information provided does not determine compliance responsibilities under OSHA standards or the *Occupational Safety and Health Act of 1970* (OSH Act).

Because interpretations and enforcement policy may change over time, you should consult the agency for the most up-to-date information. Much of it is available at the OSHA website at www.osha.gov. The website also includes locations and phone numbers for OSHA offices around the country. If you do not have access to the website, call 1-800-321-OSHA (6742). This information is available to sensory-impaired individuals upon request. Voice phone: (202) 693- 1999; teletypewriter (TTY) number: (877) 889-5627.

A MESSAGE FROM DR. DAVID MICHAELS

In 1970, the United States Congress and President Richard Nixon created the Occupational Safety and Health Administration (OSHA), a national public health agency dedicated to the basic proposition that no worker should have to choose between their life and their job.

* This is an edited, reformatted and augmented version of a publication, OSHA 3302-09R, issued by the U.S. Department of Labor, 2015.

Passed with bipartisan support, the creation of OSHA was a historic moment of cooperative national reform. The OSHA law makes it clear that the right to a safe workplace is a basic human right.

Since OSHA's first day on the job, the agency has delivered remarkable progress for our nation. Workplace injuries, illnesses and deaths have fallen dramatically. Together with our state partners, OSHA has tackled deadly safety hazards and health risks. We have established common sense standards and enforced the law against those who put workers at risk. Our standards, enforcement actions, compliance assistance and cooperative programs have saved thousands of lives and prevented countless injuries and illnesses.

Looking to the future, OSHA is committed to protecting workers from toxic chemicals and deadly safety hazards at work, ensuring that vulnerable workers in high-risk jobs have access to critical information and education about job hazards, and providing employers with vigorous compliance assistance to promote best practices that can save lives.

Although our task is far from complete, our progress gives us hope and confidence that OSHA will continue to make a lasting difference in the lives of our nation's 130 million workers, their families and their communities.

David Michaels, PhD, MPH
Assistant Secretary of Labor for OSHA

OSHA's Mission

Congress created OSHA to assure safe and healthful conditions for working men and women by setting and enforcing standards and providing training, outreach, education and compliance assistance.

Under the OSHA law, employers are responsible for providing a safe and healthful workplace for their workers. For more information, visit OSHA's website at www.osha.gov.

Introduction

On December 29, 1970, President Nixon signed the *Occupational Safety and Health Act of 1970* (OSH Act) into law, establishing OSHA. Coupled with the efforts of employers, workers, safety and health professionals, unions and advocates, OSHA and its state partners have dramatically improved workplace safety, reducing work-related deaths and injuries by more than 65 percent.

In 1970, an estimated 14,000 workers were killed on the job – about 38 every day. For 2010, the Bureau of Labor Statistics reports this number fell to about 4,500 or about 12 workers per day. At the same time, U.S. employment has almost doubled to over 130 million workers at more than 7.2 million worksites. The rate of reported serious workplace injuries and illnesses has also dropped markedly, from 11 per 100 workers in 1972 to 3.5 per 100 workers in 2010.

OSHA's safety and health standards, including those for asbestos, fall protection, cotton dust, trenching, machine guarding, benzene, lead and bloodborne pathogens have prevented countless work-related injuries, illnesses and deaths. Nevertheless, far too many preventable injuries and fatalities continue to occur. Significant hazards and unsafe conditions still exist in U.S. workplaces; each year more than 3.3 million working men and women suffer a serious job-related injury or illness. Millions more are exposed to toxic chemicals that may cause illnesses years from now.

In addition to the direct impact on individual workers, the negative consequences for America's economy are substantial. Occupational injuries and illnesses cost American employers more than $53 billion a year – over $1 billion a week – in workers' compensation costs alone. Indirect costs to employers, including lost productivity, employee training and replacement costs, and time for investigations following injuries can more than double these costs. Workers and their families suffer great emotional and psychological costs, in addition to the loss of wages and the costs of caring for the injured, which further weakens the economy.

OSHA COVERAGE

The OSH Act covers most private sector employers and their workers, in addition to some public sector employers and workers in the 50 states and certain territories and jurisdictions under federal authority. Those jurisdictions include the District of Columbia, Puerto Rico, the Virgin Islands, American Samoa, Guam, Northern Mariana Islands, Wake Island, Johnston Island, and the Outer Continental Shelf Lands as defined in the *Outer Continental Shelf Lands Act.*

Private Sector Workers

OSHA covers most private sector employers and workers in all 50 states, the District of Columbia, and other U.S. jurisdictions either directly through Federal OSHA or through an OSHA-approved state plan.

State plans are OSHA-approved job safety and health programs operated by individual states instead of Federal OSHA. The OSH Act encourages states to develop and operate their own job safety and health programs and precludes state enforcement of OSHA standards unless the state has an approved program. OSHA approves and monitors all state plans and provides as much as fifty percent of the funding for each program. State-run safety and health programs must be at least as effective as the Federal OSHA program. To find the contact information for the OSHA Federal or state plan office nearest you, call 1-800- 321-OSHA (6742) or go to www.osha.gov.

The following 22 states or territories have OSHA-approved state programs:

• Alaska	• Arizona
• California	• Hawaii
• Indiana	• Iowa
• Kentucky	• Maryland
• Michigan	• Minnesota
• Nevada	• New Mexico
• North Carolina	• Oregon
• Puerto Rico	• South Carolina
• Tennessee	• Utah
• Vermont	• Virginia
• Washington	• Wyoming

Federal OSHA provides coverage to certain workers specifically excluded from a state's plan, for example, those in some states who work in maritime industries or on military bases.

Any interested person or group, including individual workers, with a complaint concerning the operation or administration of a state program may submit a complaint to the appropriate Federal OSHA regional administrator (regional offices are listed at the end of this guide). This is called a Complaint About State Program Administration (CASPA). The complainant's name will be kept confidential. The OSHA regional administrator will investigate all such complaints, and where complaints are found to be valid, may require appropriate corrective action on the part of the state.

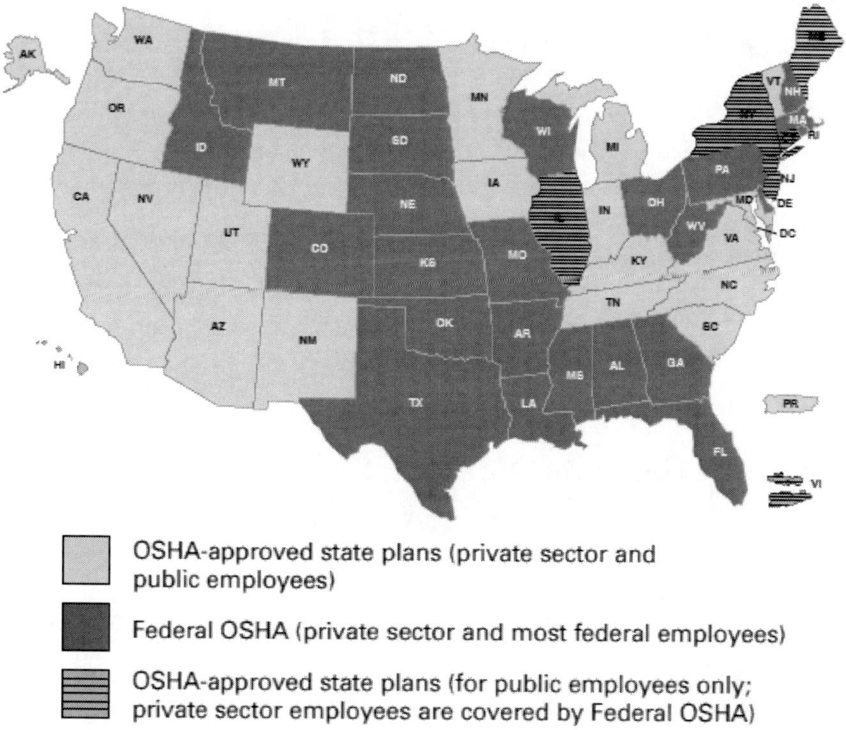

Figure. OSHA-Approved State Plans.

State and Local Government Workers

Workers at state and local government agencies are not covered by Federal OSHA, but have OSH Act protections if they work in those states that have an OSHA-approved state program.

OSHA rules also permit states and territories to develop plans that cover only public sector (state and local government) workers. In these cases, private sector workers and employers remain under Federal OSHA jurisdiction. Five additional states and one U.S. territory have OSHA-approved state plans that cover public sector workers only:

• Connecticut	• Illinois	• Maine
• New Jersey	• New York	• Virgin Islands

Federal Government Workers

OSHA's protection applies to all federal agencies. Section 19 of the OSH Act makes federal agency heads responsible for providing safe and healthful working conditions for their workers. Although OSHA does not fine federal agencies, it does monitor these agencies and conducts federal workplace inspections in response to workers' reports of hazards.

Federal agencies must have a safety and health program that meets the same standards as private employers. Under a 1998 amendment, the OSH Act covers the U.S. Postal Service the same as any private sector employer.

Not Covered under the OSH Act

- The self-employed;
- Immediate family members of farm employers; and
- Workplace hazards regulated by another federal agency (for example, the Mine Safety and Health Administration, the Department of Energy, or Coast Guard).

Rights and Responsibilities under OSHA Law

Employers have the responsibility to provide a safe workplace. **Employers MUST provide their workers with a workplace that does not have serious hazards and must follow all OSHA safety and health standards.** Employers must find and correct safety and health problems. OSHA further requires that employers must first try to eliminate or reduce hazards by making feasible changes in working conditions rather than relying on personal protective equipment such as masks, gloves, or earplugs. Switching to safer chemicals, enclosing processes to trap harmful fumes, or using ventilation systems to clean the air are examples of effective ways to eliminate or reduce risks.

Employers MUST also:

- Prominently display the official OSHA *Job Safety and Health – It's the Law* poster that describes rights and responsibilities under the OSH Act. **This poster is free and can be downloaded from www.osha.gov.**
- Inform workers about chemical hazards through training, labels, alarms, color-coded systems, chemical information sheets and other methods.
- Provide safety training to workers in a language and vocabulary they can understand.
- Keep accurate records of work-related injuries and illnesses.
- Perform tests in the workplace, such as air sampling, required by some OSHA standards.
- Provide required personal protective equipment at no cost to workers.*
- Provide hearing exams or other medical tests required by OSHA standards.
- Post OSHA citations and injury and illness data where workers can see them.
- Notify OSHA within 8 hours of a workplace fatality or within 24 hours of any work-related inpatient hospitalization, amputation or loss of an eye (1-800-321-OSHA [6742]).
- Not retaliate against workers for using their rights under the law, including their right to report a work-related injury or illness.

Under OSHA law, workers are entitled to working conditions that do not pose a risk of serious harm.

Workers have the right to:

- Working conditions that do not pose a risk of serious harm.
- File a confidential complaint with OSHA to have their workplace inspected.
- Receive information and training about hazards, methods to prevent harm, and the OSHA standards that apply to their workplace. The

* Employers must pay for most types of required personal protective equipment.

training must be done in a language and vocabulary workers can understand.

- Receive copies of records of work-related injuries and illnesses that occur in their workplace.
- Receive copies of the results from tests and monitoring done to find and measure hazards in their workplace.
- Receive copies of their workplace medical records.
- Participate in an OSHA inspection and speak in private with the inspector.
- File a complaint with OSHA if they have been retaliated against by their employer as the result of requesting an inspection or using any of their other rights under the OSH Act.
- File a complaint if punished or retaliated against for acting as a "whistleblower" under the 21 additional federal laws for which OSHA has jurisdiction.

For more information, visit OSHA's Workers' Rights page at www.osha.gov/workers.

OSHA Standards

OSHA's Construction, General Industry, Maritime and Agriculture standards protect workers from a wide range of serious hazards. Examples of OSHA standards include requirements for employers to:

- provide fall protection;
- prevent trenching cave-ins;
- prevent exposure to some infectious diseases;
- ensure the safety of workers who enter confined spaces;
- prevent exposure to harmful chemicals;
- put guards on dangerous machines;
- provide respirators or other safety equipment; and
- provide training for certain dangerous jobs in a language and vocabulary workers can understand.

Employers must also comply with the General Duty Clause of the OSH Act. This clause requires employers to keep their workplaces free of serious recognized hazards and is generally cited when no specific OSHA standard applies to the hazard.

The Standards-Setting Process

OSHA has the authority to issue new or revised occupational safety and health standards. The OSHA standards-setting process involves many steps and provides many opportunities for public engagement. OSHA can begin standards-setting procedures on its own initiative or in response to recommendations or petitions from other parties, including:

- The National Institute for Occupational Safety and Health (NIOSH), the research agency for occupational safety and health. (For more information, call 1-800-CDC-INFO (1-800-232-4636) or visit the agency's website at www.cdc.gov/niosh);
- State and local governments;
- Nationally recognized standards-producing organizations;
- Employer or labor representatives; and
- Any other interested parties.

When OSHA is considering whether to develop a new or revised standard, the Agency often publishes a Request for Information (RFI) or an Advance Notice of Proposed Rulemaking (ANPRM) in the *Federal Register* to obtain information and views from interested members of the public. OSHA will also frequently hold stakeholder meetings with interested parties to solicit information and opinions on how the Agency should proceed with the regulation. When OSHA publishes an RFI or ANPRM, interested parties can submit written comments at www.regulations.gov, where all information and submissions are made public.

If OSHA decides to proceed with issuing a new or revised regulation, it must first publish a Notice of Proposed Rulemaking (NPRM) in the *Federal Register* and solicit public comment. The NPRM contains a proposed standard along with OSHA's explanation of the need for the various requirements in that proposed standard.

Interested parties are invited to submit written comments through www.regulations.gov, and OSHA will often hold public hearings in which stakeholders can offer testimony and provide information to assist the Agency in developing a final standard. After considering all of the information and testimony provided, OSHA develops and issues a final standard that becomes enforceable.

Each spring and fall, the Department of Labor publishes in the *Federal Register* a list of all regulatory projects underway. The Regulatory Agenda provides a projected schedule for these projects to inform stakeholders of the Agency's regulatory priorities and enable interested parties to take advantage of opportunities to participate in the regulatory process. Current and past issues of the Regulatory Agenda can be accessed on OSHA's Law and Regulations page at www.osha. gov/law-regs.html.

Input from Small Business

The *Small Business Regulatory Enforcement Fairness Act of 1996* (SBREFA) gives small businesses help in understanding and complying with OSHA regulations and allows them a voice in developing new regulations. Under SBREFA, OSHA must:

- Produce Small Entity Compliance Guides for some agency rules;
- Be responsive to small business inquiries about complying with the Agency's regulations;
- Have a penalty reduction policy for small businesses;
- Involve small businesses in developing proposed rules expected to significantly affect a large number of small entities through Small Business Advocacy Review Panels; and
- Give small businesses the opportunity to challenge in court agency rules or regulations that they believe will adversely affect them.

More information about OSHA standards and the standards-setting process is available on OSHA's website at www.osha.gov. Standards can be viewed on OSHA's Law and Regulations page at www.osha.gov/law-regs.html.

Enforcement

OSHA Inspection Activities: Carrying Out Our Mission

Enforcement plays an important part in OSHA's efforts to reduce workplace injuries, illnesses, and fatalities. When OSHA finds employers who fail to uphold their safety and health responsibilities, the agency takes strong, decisive actions.

Inspections are initiated without advance notice, conducted using on-site or telephone and facsimile investigations, performed by highly trained compliance officers and scheduled based on the following priorities:

- Imminent danger;
- Catastrophes – fatalities or hospitalizations;
- Worker complaints and referrals;
- Targeted inspections – particular hazards, high injury rates; and
- Follow-up inspections.

Current workers or their representatives may file a written complaint and ask OSHA to inspect their workplace if they believe there is a serious hazard or that their employer is not following OSHA standards. Workers and their representatives have the right to ask for an inspection without OSHA telling their employer who filed the complaint. It is a violation of the OSH Act for an employer to fire, demote, transfer or in any way retaliate against a worker for filing a complaint or using other OSHA rights.

The on-site inspection begins with the presentation of the compliance officer's credentials. The compliance officer will explain why OSHA selected the workplace for inspection and describe the scope of the inspection process, walkaround procedures, employee representation and employee interviews. Following the opening conference, the compliance officer and the representatives will walk through portions of the workplace covered by the inspection, inspecting for hazards that could lead to worker injury or illness. After the walkaround, the compliance officer will hold a closing conference with the employer and the employee representative to discuss the findings.

When an inspector finds violations of OSHA standards or serious hazards, OSHA may issue citations and fines. A citation includes methods an employer may use to fix a problem and the date by which the corrective actions must be completed.

Employers have the right to contest any part of the citation, including whether a violation actually exists. Workers only have the right to challenge the deadline by which a problem must be resolved. Appeals of citations are heard by the independent Occupational Safety and Health Review Commission (OSHRC). To contact the OSHRC, visit www.oshrc.gov or call (202) 606-5370.

OSHA carries out its enforcement activities through its 10 regional offices and 90 area offices. OSHA's regional offices are located in Boston, New York City, Philadelphia, Atlanta, Chicago, Dallas, Kansas City, Denver, San Francisco and Seattle. Contact information for each regional office is available at the end of this guide.

Severe Violator Enforcement Program

OSHA's Severe Violator Enforcement Program (SVEP) became effective on June 18, 2010. The program focuses enforcement efforts on employers who willfully and repeatedly endanger workers by exposing them to serious hazards. The SVEP directive establishes procedures and enforcement actions for these violators, including mandatory follow-up inspections of workplaces found in violation and inspections of other worksites of the same company where similar hazards or deficiencies may be present. Visit www.osha.gov for more information.

General Reporting and Recordkeeping Requirements

OSHA's Reporting Requirements

All employers must report to OSHA:

- The death of any worker from a work-related incident within 8 hours of learning about it;
- All work-related inpatient hospitalizations, amputations and losses of an eye within 24 hours.

For more information, visit www.osha.gov/ recordkeeping.

In addition, employers must report all fatal heart attacks that occur at work. Deaths from motor vehicle accidents on public streets (except those in a construction work zone) and in accidents on commercial airplanes, trains, subways or buses do not need to be reported.

These reports may be made by telephone or in person to the nearest OSHA area office listed at www.osha.gov or by calling OSHA's toll-free number, 1-800-321-OSHA (6742).

OSHA's Recordkeeping Requirements

Tracking and investigating workplace injuries and illnesses play an important role in preventing future injuries and illnesses, and for that reason, OSHA requires certain covered employers in high-hazard industries to prepare and maintain records of serious work-related injuries and illnesses.

Employers with more than ten employees and whose establishments are not classified as a partially exempt industry must record serious work-related injuries and illnesses using OSHA Forms 300, 300A and 301, which are available at www.osha.gov/recordkeeping/RKforms. html. A list of partially exempt industries, including establishments in specific low hazard retail, service, finance, insurance or real estate industries is available at www.osha.gov/ recordkeeping/ppt1/RK1exempttable.html.

Employers who are required to keep Form 300, the Injury and Illness log, must also post Form 300A, the Summary of Work-Related Injuries and Illnesses, in the workplace every year from February 1 to April 30. For more information, visit www.osha.gov/recordkeeping.

Employers and workers need accurate, timely information to focus their prevention activities, and OSHA uses this information for many purposes, including inspection targeting, performance measurement, standards development and resource allocation. Injury and illness data also aid employers and workers in identifying possible safety and health hazards at the employer's establishment. OSHA encourages employers to review and investigate patterns of injuries and illnesses, and to conduct investigations of injuries and near misses to prevent similar events in the future.

OSHA is responsible for administering the recordkeeping system established by the OSH Act. OSHA's recordkeeping regulations provide specific recording and reporting requirements which comprise the framework for the nationwide occupational safety and health recordkeeping system. For more information about OSHA's recordkeeping requirements visit www.osha.gov/recordkeeping.

FILING A COMPLAINT

Hazardous Workplace Complaints

If a workplace has unsafe or unhealthful working conditions, workers may want to file a complaint. Often the best and fastest way to get a hazard corrected is to notify a supervisor or employer.

Workers or their representatives may file a complaint online or by phone, mail, email or fax with the nearest OSHA office and request an inspection. A worker may also ask OSHA not to reveal his or her name. To file a complaint, call 1-800-321-OSHA (6742) or contact the nearest OSHA regional, area, state plan, or consultation office listed at www.osha.gov. The teletypewriter (TTY) number is (877) 889-5627.

Written, Signed Complaints Submitted to OSHA Area Offices Are More Likely to Result in an on-Site OSHA Inspection

Most online or unsigned complaints are resolved informally over the phone with the employer. Complaints from workers in states with an OSHA-approved state plan will be forwarded to the appropriate state plan for response.

Workers can call 1-800-321-OSHA (6742) to request a complaint form from their local OSHA office or visit www.osha.gov/pls/osha7/eComplaintForm.html to submit the form online. Completed forms can also be faxed or mailed to the local OSHA office (provided at the end of this guide). Include your name, address and telephone number so that OSHA can contact you.

OSHA'S WHISTLEBLOWER PROGRAM: PROTECTION FROM RETALIATION

To help ensure that workers are free to participate in safety and health activities, Section 11(c) of the OSH Act prohibits any person from discharging or in any manner retaliating against any worker for exercising rights under the

OSH Act. These rights include raising safety and health concerns with an employer, reporting a work-related injury or illness, filing a complaint with OSHA, seeking an OSHA inspection, participating in an OSHA inspection and participating or testifying in any proceeding related to an OSHA inspection.

Protection from retaliation means that an employer cannot retaliate by taking "adverse action" against workers, such as:

- Firing or laying off;
- Blacklisting;
- Demoting;
- Denying overtime or promotion;
- Disciplining;
- Denying of benefits;
- Failing to hire or rehire;
- Intimidation;
- Making threats;
- Reassignment affecting prospects for promotion; or
- Reducing pay or hours.

If a worker believes an employer has retaliated against them for exercising their safety and health rights, they should contact their local OSHA office right away. You must file a retaliation complaint with OSHA within 30 calendar days from the date the retaliatory decision has been both made and communicated to the worker. **No form is needed, but workers must call OSHA within 30 days of the alleged retaliation** (at 1-800-321-OSHA [6742]). For more information, please visit www.whistleblowers.gov.

If There is a Dangerous Situation at Work

If a worker believes working conditions are unsafe or unhealthful, OSHA recommends that he or she bring the conditions to the employer's attention, if possible. A worker may file a complaint with OSHA concerning a hazardous working condition at any time. However, workers should not leave the worksite merely because they have filed a complaint. If the condition clearly presents a risk of death or serious physical harm, there is not sufficient time for OSHA to inspect, and, where possible, a worker has brought the condition to the attention of the employer, the worker may have a legal right to refuse to work in a situation in which he or she would be exposed to the hazard.

If a worker, with no reasonable alternative, refuses in good faith to expose himself or herself to a dangerous condition, he or she would be protected from subsequent retaliation. The condition must be of such a nature that a reasonable person would conclude that there is a real danger of death or serious harm and that there is not enough time to contact OSHA and for OSHA to inspect. Where possible, the worker must have also sought from his or her employer, and been unable to obtain, a correction of the condition. For more information, go to www.osha.gov/workers.

ADDITIONAL WHISTLEBLOWER PROTECTIONS

Since passage of the OSH Act in 1970, Congress has expanded OSHA's whistleblower protection authority to protect workers from retaliation under a total of 22 federal laws. These laws protect workers who report violations of various workplace safety, airline, commercial motor carrier, consumer product, environmental, financial reform, healthcare reform, nuclear, pipeline, public transportation agency, railroad, maritime and securities laws. Complaints must be reported to OSHA within set timeframes following the retaliatory action, as prescribed by each law. These laws, and the number of days workers have to file a complaint, are:

Worker, Environmental and Nuclear Safety Laws

- ***Asbestos Hazard Emergency Response Act (AHERA)*** (90 days). Provides retaliation protection for individuals who report violations of environmental laws relating to asbestos in public or private nonprofit elementary and secondary school systems.
- ***Clean Air Act (CAA)*** (30 days). Provides retaliation protection for employees who, among other things, report violations of this law, which provides for the development and enforcement of standards regarding air quality and air pollution.
- ***Comprehensive Environmental Response, Compensation, and Liability Act (CERCLA)*** (30 days). Protects employees who report regulatory violations involving accidents, spills, and other emergency

releases of pollutants into the environment. The law also protects employees who report violations related to the cleanup of uncontrolled or abandoned hazardous waste sites.

- ***Energy Reorganization Act (ERA)*** (180 days). Protects certain employees in the nuclear industry who report violations of the *Atomic Energy Act (AEA).* Protected employees include employees of operators, contractors and subcontractors of nuclear power plants licensed by the Nuclear Regulatory Commission, and employees of contractors working with the Department of Energy under a contract pursuant to the *Atomic Energy Act.*
- ***Federal Water Pollution Control Act (FWPCA) (also known as the Clean Water Act)*** (30 days). Provides retaliation protection for employees who, among other things, report violations of the law controlling water pollution.
- ***Occupational Safety and Health Act of 1970*** (30 days). Provides retaliation protection for employees who exercise a variety of rights guaranteed under this law, such as filing a safety and health complaint with OSHA and participating in an inspection.
- ***Safe Drinking Water Act (SDWA)*** (30 days). Provides retaliation protection for employees who, among other things, report violations of this law, which requires that all drinking water systems assure that their water is potable, as determined by the Environmental Protection Agency.
- ***Solid Waste Disposal Act (SWDA) (also known as the Resource Conservation and Recovery Act)*** (30 days). Provides retaliation protection for employees who, among other things, report violations of the law regulating the disposal of solid waste.
- ***Toxic Substances Control Act (TSCA)*** (30 days). Provides retaliation protection for employees who, among other things, report violations of regulations involving the manufacture, distribution, and use of certain toxic substances.

Transportation Industry Laws

- ***Federal Railroad Safety Act (FRSA)*** (180 days). Provides protection to employees of railroad carriers and contractors and subcontractors of those carriers who report an alleged violation of any federal law, rule, or regulation relating to railroad safety or security, or gross fraud, waste, or abuse of federal grants or other public funds intended to be used for railroad safety or security; report, in good faith, a hazardous safety or security condition; refuse to violate or assist in the violation of any federal law, rule, or regulation relating to railroad safety or security; refuse to work when confronted by a hazardous safety or security condition related to the performance of the employee's duties (under imminent danger circumstances); request prompt medical or first-aid treatment for employment-related injuries; are disciplined for requesting medical or first-aid treatment or for following an order or treatment plan of a treating physician.

- ***International Safe Container Act (ISCA)*** (60 days). Provides retaliation protection for employees who report violations of this law, which regulates shipping containers.
- ***Moving Ahead for Progress in the 21st Century Act (MAP-21)*** (180 days). Prohibits retaliation by motor vehicle manufacturers, part suppliers, and dealerships against employees for providing information to the employer or the U.S. Department of Transportation about motor vehicle defects, noncompliance, or violations of the notification or reporting requirements enforced by the National Highway Traffic Safety Administration or for engaging in related protected activities as set forth in the provision.
- ***National Transit Systems Security Act (NTSSA)*** (180 days). Provides protection to public transit employees who, among other things, report an alleged violation of any federal law, rule, or regulation relating to public transportation agency safety or security, or fraud, waste, or abuse of federal grants or other public funds intended to be used for public transportation safety or security; refuse to violate or assist in the violation of any federal law, rule, or regulation relating to public transportation safety or security; report a hazardous safety or security condition; refuse to work when confronted by a hazardous safety or security condition related to the performance of the employee's duties (under imminent danger circumstances).
- ***Pipeline Safety Improvement Act of 2002 (PSIA)*** (180 days). Provides retaliation protection for employees who report violations of the federal laws regarding pipeline safety and security or who refuse to violate such provisions.
- ***Seaman's Protection Act (SPA)*** (180 days). Seamen are protected, among other things, for reporting to the Coast Guard or other federal

agency a reasonably believed violation of a maritime safety law or regulation prescribed under that law or regulation. The law also protects work refusals where the employee reasonably believes an assigned task would result in serious injury or impairment of health to the seaman, other seamen, or the public and when the seaman sought, and was unable to obtain correction of the unsafe conditions.

- ***Surface Transportation Assistance Act (STAA)*** (180 days). Provides retaliation protection for truck drivers and other employees relating to the safety of commercial motor vehicles. Coverage includes all buses for hire and freight trucks with a gross vehicle weight greater than 10,001 pounds.
- ***Wendell H. Ford Aviation Investment and Reform Act for the 21st Century (AIR21)*** (90 days). Provides retaliation protection for employees of air carriers, contractors, or subcontractors of air carriers who, among other things, raise safety concerns.

Fraud Prevention Laws

- ***Affordable Care Act (ACA)*** (180 days). Protects employees who report violations of any provision of Title I of the ACA, including but not limited to retaliation based on an individual's receipt of health insurance subsidies, the denial of coverage based on a preexisting condition, or an insurer's failure to rebate a portion of an excess premium.
- ***Consumer Financial Protection Act of 2010 (CFPA), Section 1057 of the Dodd-Frank Wall Street Reform and Consumer Protection Act*** (180 days). Protects employees who report perceived violations of any provision of the *Dodd-Frank Act,* which encompasses nearly every aspect of the financial services industry. The law also protects employees who report violations of any rule, order, standard or prohibition prescribed by the Bureau of Consumer Financial Protection.
- ***Section 806 of the Sarbanes-Oxley Act of 2002 (SOX)*** (180 days). Protects employees of certain companies who report alleged mail, wire, bank or securities fraud; violations of the Securities and Exchange Commission (SEC) rules and regulations; or violations of Federal laws related to fraud against shareholders. The law covers

employees of publicly traded companies and companies required to file certain reports with the SEC.

Consumer Safety Laws

- ***Consumer Product Safety Improvement Act (CPSIA)*** (180 days). Protects employees who report to their employer, the federal government, or a state attorney general reasonably perceived violations of any statute or regulation within the jurisdiction of the Consumer Product Safety Commission (CPSC). CPSIA covers employees of consumer product manufacturers, importers, distributors, retailers, and private labelers.
- ***FDA Food Safety Modernization Act (FSMA)*** (180 days). Protects employees of food manufacturers, distributors, packers, and transporters for reporting a violation of the *Food, Drug, and Cosmetic Act,* or a regulation promulgated under this law. Employees are also protected from retaliation for refusing to participate in a practice that violates this law.

If you believe that you have been retaliated against, call 1-800-321-OSHA (6742) to be connected to the nearest OSHA office to report your complaint. For more information, visit OSHA's Whistleblower page at www.whistleblowers.gov.

OSHA Assistance, Services and Programs

OSHA has a great deal of information to assist employers in complying with their responsibilities under OSHA law. Several OSHA programs and services can help employers identify and correct job hazards, as well as improve their injury and illness prevention program.

Establishing an Injury and Illness Prevention Program

The key to a safe and healthful work environment is a comprehensive injury and illness prevention program.

Injury and illness prevention programs are systems that can substantially reduce the number and severity of workplace injuries and illnesses, while reducing costs to employers. Thousands of employers across the United States already manage safety using injury and illness prevention programs, and OSHA believes that all employers can and should do the same. Thirty-four states have requirements or voluntary guidelines for workplace injury and illness prevention programs. Most successful injury and illness prevention programs are based on a common set of key elements. These include management leadership, worker participation, hazard identification, hazard prevention and control, education and training, and program evaluation and improvement. Visit OSHA's Injury and Illness Prevention Programs web page at www.osha.gov/dsg/topics/safetyhealth for more information.

Compliance Assistance Specialists

OSHA has compliance assistance specialists throughout the nation located in most OSHA offices. Compliance assistance specialists can provide information to employers and workers about OSHA standards, short educational programs on specific hazards or OSHA rights and responsibilities, and information on additional compliance assistance resources. For more details, visit www.osha.gov/dcsp/compliance_ assistance/cas.html or call 1-800-321-OSHA (6742) to contact your local OSHA office.

Free on-site Safety and Health Consultation Services for Small Business

OSHA's On-site Consultation Program offers free and confidential advice to small and medium-sized businesses in all states across the country, with priority given to high-hazard worksites. Each year, responding to requests from small employers looking to create or improve their safety and health management programs, OSHA's On-site Consultation Program conducts over 29,000 visits to small business worksites covering over 1.5 million workers across the nation.

On-site consultation services are separate from enforcement and do not result in penalties or citations. Consultants from state agencies or universities work with employers to identify workplace hazards, provide advice on compliance with OSHA standards, and assist in establishing safety and health management programs.

For more information, to find the local On-site Consultation office in your state, or to request a brochure on consultation services, visit www.osha.gov/consultation, or call 1-800-321-OSHA (6742).

Under the consultation program, certain exemplary employers may request participation in OSHA's **Safety and Health Achievement Recognition Program (SHARP)**. Eligibility for participation includes, but is not limited to, receiving a full-service, comprehensive consultation visit, correcting all identified hazards and developing an effective safety and health management program.

Worksites that receive SHARP recognition are exempt from programmed inspections during the period that the SHARP certification is valid.

Cooperative Programs

OSHA offers cooperative programs under which businesses, labor groups and other organizations can work cooperatively with OSHA. To find out more about any of the following programs, visit www.osha.gov/dcsp/compliance_assistance/index_programs.html.

Strategic Partnerships and Alliances

The OSHA Strategic Partnerships (OSP) provide the opportunity for OSHA to partner with employers, workers, professional or trade associations, labor organizations, and/or other interested stakeholders. OSHA Partnerships are formalized through unique agreements designed to encourage, assist, and recognize partner efforts to eliminate serious hazards and achieve model workplace safety and health practices. Through the Alliance Program, OSHA works with groups committed to worker safety and health to prevent workplace fatalities, injuries and illnesses by developing compliance assistance tools and resources to share with workers and employers, and educate workers and employers about their rights and responsibilities.

Voluntary Protection Programs (VPP)

The VPP recognize employers and workers in private industry and federal agencies who have implemented effective safety and health management programs and maintain injury and illness rates below the national average for their respective industries. In VPP, management, labor, and OSHA work cooperatively and proactively to prevent fatalities, injuries, and illnesses through a system focused on: hazard prevention and control, worksite analysis, training, and management commitment and worker involvement.

Occupational Safety and Health Training

The OSHA Training Institute partners with 27 OSHA Training Institute Education Centers at 42 locations throughout the United States to deliver courses on OSHA standards and occupational safety and health topics to thousands of students a year. For more information on training courses, visit www.osha.gov/otiec.

OSHA Educational Materials

OSHA has many types of educational materials in English, Spanish, Vietnamese and other languages available in print or online. These include:

- Brochures/booklets;
- Fact Sheets;
- Guidance documents that provide detailed examinations of specific safety and health issues;
- Online Safety and Health Topics pages;
- Posters;
- Small, laminated QuickCardsTM that provide brief safety and health information; and
- *QuickTakes*, OSHA's free, twice-monthly online newsletter with the latest news about OSHA initiatives and products to assist employers and workers in finding and preventing workplace hazards. To sign up for *QuickTakes* visit www.osha.gov/quicktakes.

To view materials available online or for a listing of free publications, visit www.osha.gov/ publications. You can also call 1-800-321-OSHA (6742) to order publications.

Select OSHA publications are available in e-Book format. OSHA e-Books are designed to increase readability on smartphones, tablets and other mobile devices. For access, go to www.osha.gov/ebooks.

OSHA's web site also has information on job hazards and injury and illness prevention for employers and workers. To learn more about OSHA's safety and health resources online, visit www.osha.gov or www.osha.gov/html/ a-z-index.html.

OSHA Advisory Committees

OSHA sponsors advisory committees to advise the Secretary of Labor and the Assistant Secretary of Labor for Occupational Safety and Health on workplace safety and health issues.

All of OSHA's advisory committees have membership balanced between representatives of workers and employers, and most also include other qualified individuals such as government officials, safety and health professionals and members of the public. All committees accept comments from interested individuals. Transcripts and minutes of the meetings are also available to the public on the committee web pages at www.osha.gov/osha-advisory-committee.html.

The five current advisory committees are:

- The National Advisory Committee on Occupational Safety and Health (NACOSH), which advises, consults with and makes recommendations to the U.S. Secretaries of Labor and Health and Human Services (HHS) on matters regarding the OSH Act;
- The Advisory Committee on Construction Safety and Health (ACCSH), which advises the Secretary of Labor on construction safety and health standards and other matters;
- The Federal Advisory Council on Occupational Safety and Health (FACOSH), which advises the Secretary of Labor on all aspects of federal agency safety and health;
- The Maritime Advisory Committee for Occupational Safety and Health (MACOSH), which advises the Secretary of Labor on workplace safety and health programs, policies and standards in the maritime industry; and

- The Whistleblower Protection Advisory Committee (WPAC), which advises, consults with and makes recommendations to the Secretary of Labor on ways to improve the fairness, efficiency, effectiveness, and transparency of OSHA's administration of whistleblower protections.

In addition, OSHA may form short-term advisory committees to advise the agency on specific issues.

NIOSH Health Hazard Evaluation Program

Getting Help with Health Hazards

The National Institute for Occupational Safety and Health (NIOSH) is a federal agency that conducts scientific and medical research on workers' safety and health. At no cost to employers or workers, NIOSH can help identify health hazards and recommend ways to reduce or eliminate those hazards in the workplace through its Health Hazard Evaluation (HHE) Program.

Workers, union representatives and employers can request a NIOSH HHE. An HHE is often requested when there is a higher than expected rate of a disease or injury in a group of workers. These situations may be the result of an unknown cause, a new hazard, or a mixture of sources. To request a NIOSH Health Hazard Evaluation go to www.cdc.gov/niosh/hhe/request.html. To find out more, in English or Spanish, about the Health Hazard Evaluation Program:

E-mail HHERequestHelp@cdc.gov or call 800-CDCINFO (800-232-4636).

OSHA Regional Offices

Region I

Boston Regional Office
(CT*, ME*, MA, NH, RI, VT*)
JFK Federal Building, Room E340
Boston, MA 02203
(617) 565-9860 (617) 565-9827 Fax

Region II

New York Regional Office
(NJ*, NY*, PR*, VI*)
201 Varick Street, Room 670
New York, NY 10014
(212) 337-2378 (212) 337-2371 Fax

Region III

Philadelphia Regional Office
(DE, DC, MD*, PA, VA*, WV)
The Curtis Center
170 S. Independence Mall West
Suite 740 West
Philadelphia, PA 19106-3309
(215) 861-4900 (215) 861-4904 Fax

Region IV

Atlanta Regional Office
(AL, FL, GA, KY*, MS, NC*, SC*, TN*)
61 Forsyth Street, SW, Room 6T50
Atlanta, GA 30303
(678) 237-0400 (678) 237-0447 Fax

Region V

Chicago Regional Office
(IL*, IN*, MI*, MN*, OH, WI)
230 South Dearborn Street
Room 3244
Chicago, IL 60604
(312) 353-2220 (312) 353-7774 Fax

Region VI

Dallas Regional Office
(AR, LA, NM*, OK, TX)
525 Griffin Street, Room 602
Dallas, TX 75202
(972) 850-4145 (972) 850-4149 Fax
(972) 850-4150 FSO Fax

Region VII

Kansas City Regional Office
(IA*, KS, MO, NE)
Two Pershing Square Building
2300 Main Street, Suite 1010
Kansas City, MO 64108-2416
(816) 283-8745 (816) 283-0547 Fax

Region VIII

Denver Regional Office
(CO, MT, ND, SD, UT*, WY*)
Cesar Chavez Memorial Building
1244 Speer Boulevard, Suite 551
Denver, CO 80204
(720) 264-6550 (720) 264-6585 Fax

Region IX

San Francisco Regional Office
(AZ*, CA*, HI*, NV*, and American Samoa,
Guam and the Northern Mariana Islands)
90 7th Street, Suite 18100
San Francisco, CA 94103
(415) 625-2547 (415) 625-2534 Fax

Region X

Seattle Regional Office
(AK*, ID, OR*, WA*)
300 Fifth Avenue, Suite 1280
Seattle, WA 98104
(206) 757-6700 (206) 757-6705 Fax

*These states and territories operate their own OSHA-approved job safety and health plans and cover state and local government employees as well as private sector employees. The Connecticut, Illinois, Maine, New Jersey, New York and Virgin Islands programs cover public employees only. (Private sector workers in these states are covered by Federal OSHA). States with approved programs must have standards that are identical to, or at least as effective as, the Federal OSHA standards.

Note: To get contact information for OSHA area offices, OSHA-approved state plans and OSHA consultation projects, please visit us online at www.osha.gov or call us at 1-800-321-OSHA (6742).

How to Contact OSHA

For questions or to get information or advice, to report an emergency, fatality, inpatient hospitalization, amputation, or loss of an eye, or to file a confidential complaint, contact your nearest OSHA office, visit www.osha.gov or call OSHA at 1-800-321-OSHA (6742), TTY 1-877-889-5627.

In: The Occupational Safety and Health Act
Editor: Christy Knight
ISBN: 978-1-63482-637-2

Chapter 4

WORKERS' RIGHTS*

Occupational Safety and Health Administration

OCCUPATIONAL SAFETY AND HEALTH ACT OF 1970

> "To assure safe and healthful working conditions for working men and women; by authorizing enforcement of the standards developed under the Act; by assisting and encouraging the States in their efforts to assure safe and healthful working conditions; by providing for research, information, education, and training in the field of occupational safety and health..."

This publication provides a general overview of worker rights under the *Occupational Safety and Health Act* (OSH Act). This publication does not alter or determine compliance responsibilities which are set forth in OSHA standards and the OSH Act. Moreover, because interpretations and enforcement policy may change over time, for additional guidance on OSHA compliance requirements the reader should consult current administrative interpretations and decisions by the Occupational Safety and Health Review Commission and the courts.

This document, *Workers' Rights*, replaces *Employee Workplace Rights*.

---

* This is an edited, reformatted and augmented version of a publication, OSHA 3021-09R, issued by the U.S. Department of Labor, 2014.

This information will be made available to sensory-impaired individuals upon request. Voice phone: (202) 693-1999; teletypewriter (TTY) number: 1-877-889-5627.

INTRODUCTION

Worker Protection is the Law of the Land

You have the right to a safe workplace. The *Occupational Safety and Health Act of 1970* (OSH Act) was passed to prevent workers from being killed or otherwise harmed at work. The law requires employers to provide their employees with working conditions that are free of known dangers. The OSH Act created the Occupational Safety and Health Administration (OSHA), which sets and enforces protective workplace safety and health standards. OSHA also provides information, training and assistance to employers and workers.

Contact us if you have questions or want to file a complaint. We will keep your information confidential. *We are here to help you.*

Workers' Rights under the OSH Act

The OSH Act gives workers the right to safe and healthful working conditions. It is the duty of employers to provide workplaces that are free of known dangers that could harm their employees. This law also gives workers important rights to participate in activities to ensure their protection from job hazards. This booklet explains workers' rights to:

- File a confidential complaint with OSHA to have their workplace inspected.
- Receive information and training about hazards, methods to prevent harm, and the OSHA standards that apply to their workplace. The training must be done in a language and vocabulary workers can understand.

- Review records of work-related injuries and illnesses that occur in their workplace.
- Receive copies of the results from tests and monitoring done to find and measure hazards in the workplace.
- Get copies of their workplace medical records.
- Participate in an OSHA inspection and speak in private with the inspector.
- File a complaint with OSHA if they have been retaliated against by their employer as the result of requesting an inspection or using any of their other rights under the OSH Act.
- File a complaint if punished or retaliated against for acting as a "whistleblower" under the additional 21 federal statutes for which OSHA has jurisdiction.

A job must be safe or it cannot be called a good job. OSHA strives to make sure that every worker in the nation goes home unharmed at the end of the workday, the most important right of all.

Employer Responsibilities

Employers have the responsibility to provide a safe workplace. *Employers MUST provide their employees with a workplace that does not have serious hazards and must follow all OSHA safety and health standards.* Employers must find and correct safety and health problems. OSHA further requires that employers must try to eliminate or reduce hazards first by making feasible changes in working conditions – switching to safer chemicals, enclosing processes to trap harmful fumes, or using ventilation systems to clean the air are examples of effective ways to get rid of or minimize risks – rather than just relying on personal protective equipment such as masks, gloves, or earplugs.

Employers *MUST* also:

- Prominently display the official OSHA poster that describes rights and responsibilities under the OSH Act. *This poster is free and can be downloaded from* www.osha.gov.
- Inform workers about hazards through training, labels, alarms, color-coded systems, chemical information sheets and other methods.
- Train workers in a language and vocabulary they can understand.

- Keep accurate records of work-related injuries and illnesses.
- Perform tests in the workplace, such as air sampling, required by some OSHA standards.
- Provide hearing exams or other medical tests required by OSHA standards.
- Post OSHA citations and injury and illness data where workers can see them.
- As of January 1, 2015, notify OSHA within 8 hours of a workplace fatality or within 24 hours of any work-related inpatient hospitalization, amputation or loss of an eye.
- Not retaliate against workers for using their rights under the law, including their right to report a work-related injury or illness.

WHO DOES OSHA COVER

Private Sector Workers

Most employees in the nation come under OSHA's jurisdiction. OSHA covers most private sector employers and employees in all 50 states, the District of Columbia, and other U.S. jurisdictions either directly through Federal OSHA or through an OSHA-approved state plan. State-run health and safety plans must be at least as effective as the Federal OSHA program. To find the contact information for the OSHA Federal or State Program office nearest you, call 1-800-321-OSHA (6742) or go to www.osha.gov.

State and Local Government Workers

Employees who work for state and local governments are not covered by Federal OSHA, but have OSH Act protections if they work in those states that have an OSHA-approved state plan. The following 22 states or territories have OSHA-approved programs:

Alaska	Arizona	California
Hawaii	Indiana	Iowa
Kentucky	Maryland	Michigan
Minnesota	Nevada	New Mexico

North Carolina	Oregon	South Carolina
Tennessee	Utah	Vermont
Virginia	Washington	Wyoming
Puerto Rico		

Four additional states and one U.S. territory have OSHA-approved plans that cover public sector workers only:

Connecticut	Illinois	New Jersey
New York	Virgin Islands	

Private sector workers in these four states and the Virgin Islands are covered by Federal OSHA.

Federal Government Workers

Federal agencies must have a safety and health program that meets the same standards as private employers. Although OSHA does not fine federal agencies, it does monitor federal agencies and responds to workers' complaints. The United States Postal Service (USPS) is covered by OSHA.

Not Covered under the OSH Act

- Self-employed;
- Immediate family members of farm employers; and
- Workplace hazards regulated by another federal agency (for example, the Mine Safety and Health Administration, the Department of Energy, or Coast Guard).

Worker Rights in State-Plan States

States that assume responsibility for their own occupational safety and health programs must have provisions at least as effective as Federal OSHA's, including the protection of worker rights.

OSHA-Approved State Plans

Any interested person or group, including employees, with a complaint concerning the operation or administration of a state program may submit a complaint to the appropriate Federal OSHA regional administrator. (See contact list at the end of this booklet). This is called a Complaint About State Program Administration (CASPA). The complainant's name will be kept confidential. The OSHA regional administrator will investigate all such complaints, and where complaints are found to be valid, require appropriate corrective action on the part of the state.

RIGHT TO A SAFE AND HEALTHFUL WORKPLACE

Employers' "General Duty"

Employers have the responsibility to provide a safe and healthful workplace that is free from serious recognized hazards. This is commonly known as the General Duty Clause of the OSH Act.

OSHA Standards: Protection on the Job OSHA standards are rules that describe the methods that employers must use to protect their employees from hazards. There are four groups of OSHA standards: General Industry, Construction, Maritime, and Agriculture. (General Industry is the set that applies to the largest number of workers and worksites). These standards are designed to protect workers from a wide range of hazards.

These standards also limit the amount of hazardous chemicals, substances, or noise that workers can be exposed to; require the use of certain safe work practices and equipment; and require employers to monitor certain hazards and keep records of workplace injuries and illnesses.

Examples of OSHA standards include requirements to:

- Provide fall protection, such as a safety harness and lifeline;
- Prevent trenching cave-ins;
- Ensure the safety of workers who enter confined spaces such as manholes or grain bins;
- Prevent exposure to high levels of noise that can damage hearing;
- Put guards on machines;
- Prevent exposure to harmful levels of substances like asbestos and lead;
- Provide workers with respirators and other needed safety equipment (in almost all cases, free of charge);
- Provide healthcare workers with needles and sharp instruments that have built-in safety features to prevent skin punctures or cuts that could cause exposure to infectious diseases; and
- Train workers using a language and vocabulary they understand about hazards and how to protect themselves.

Employers must also comply with the General Duty Clause of the OSH Act. This clause requires employers to keep their workplaces free of serious recognized hazards and is generally cited when no specific OSHA standard applies to the hazard.

Right to be Provided Protective Equipment Free of Charge

In some situations it is not possible to completely eliminate a hazard or reduce exposures to a safe level, so respirators, goggles, earplugs, gloves, or other types of personal protective equipment are often used by themselves or in addition to other hazard control measures. Employers must provide most protective equipment free of charge. Employers are responsible for knowing when protective equipment is needed.

Right to Information

OSHA gives workers and their representatives the right to see information that employers collect on hazards in the workplace. Workers have the right to know what hazards are present in the workplace and how to protect themselves. Many OSHA standards require various methods that employers must use to inform their employees, such as warning signs, color-coding, signals, and training. Workers must receive their normal rate of pay to attend training that is required by OSHA standards and rules. The training must be in a language and vocabulary that workers can understand.

RIGHT TO KNOW ABOUT CHEMICAL HAZARDS

The Hazard Communication standard, known as the "right-to-know" standard, requires employers to inform and train workers about hazardous chemicals and substances in the workplace. Employers must:

- Provide workers with effective information and training on hazardous chemicals in their work area.

This training must be in a language and vocabulary that workers can understand;

- n Keep a current list of hazardous chemicals that are in the workplace;
- n Make sure that hazardous chemical containers are properly labeled with the identity of the hazardous chemical and appropriate hazard warnings; and
- n Have and make available to workers and their representatives Safety Data Sheets (SDSs) (formerly known as Material Safety Data Sheets or MSDSs) for each substance that provide detailed information about

chemical hazards, their effects, how to prevent exposure, and emergency treatment if an exposure occurs.

RIGHT TO KNOW ABOUT LAWS AND YOUR RIGHTS

Employers must display the official OSHA Poster, *Job Safety and Health: It's the Law*, in a place where workers will see it. It can be downloaded from the OSHA website, www.osha.gov. Pre-printed copies can also be obtained from OSHA.

RIGHT TO GET COPIES OF WORKPLACE INJURY AND ILLNESS RECORDS

OSHA's Recordkeeping Rule requires employers in higher-hazard industries with more than ten employees to keep accurate and complete records of work-related injuries and illnesses. (Certain low-hazard workplaces such as offices are not required to keep such records). Employers must record any serious work-related injury or illness on the OSHA Form 300. A serious injury or illness is one that required medical treatment other than first aid, restricted work or days away from work. (Details of each incident are entered on a separate form, the OSHA Form 301). This OSHA Form 300 becomes an ongoing log of all recordable incidents. Each year from February 1 through April 30, employers must post a summary of the injury and illness log from the previous year (OSHA Form 300A) in a place where workers can see it. Workers and their representatives have the right to receive copies of the full OSHA Form 300 log. Following a request, employers must make copies available at the end of the next business day.

These injury and illness logs are important because they provide a comprehensive guide to possible hazards in the workplace that may need correcting. The logs should be used to focus on areas with high injury and illness rates, and to find and fix hazards in order to prevent future occurrences.

RIGHT TO EXPOSURE DATA

Many OSHA standards require employers to run tests of the workplace environment to find out if their workers are being exposed to harmful levels of hazardous substances such as lead or asbestos, or high levels of noise or radiation. These types of tests are called exposure monitoring. OSHA gives workers the right to get the results of these tests.

RIGHT TO YOUR MEDICAL RECORDS

Some OSHA standards require medical tests to find out if a worker's health has been affected because of exposures at work. For example, employers must test for hearing loss in workers exposed to excessive noise or for decreased lung function in workers exposed to asbestos. Workers have a right to their medical records. Workers' representatives also have a right to review these records but they must first get written permission from the worker to gain access to their medical information.

OSHA WORKSITE INVESTIGATIONS

OSHA conducts on-site inspections of worksites to enforce the OSHA law that protects workers and their rights. Inspections are initiated without advance notice, conducted using on-site or telephone and facsimile investigations, and performed by highly trained compliance officers. Worksite inspections are conducted based on the following priorities:

- Imminent danger;
- A fatality or hospitalizations; n Worker complaints and referrals;
- Targeted inspections – particular hazards, high injury rates; and
- Follow-up inspections.

Inspections are conducted without employers knowing when or where they will occur. The employer is not informed in advance that there will be an inspection, regardless of whether it is in response to a complaint or is a programmed inspection.

Right to File a Complaint with OSHA to Request an On-site OSHA Inspection

On-site inspections can be triggered by a worker complaint of a potential workplace hazard or violation. If your workplace has unsafe or unhealthful working conditions, you may want to file a complaint. Often the best and fastest way to get a hazard corrected is to notify your supervisor or employer.

Current workers or their representatives may file a written complaint and ask OSHA to inspect their workplace if they believe there is a serious hazard or that their employer is not following OSHA standards or rules. *Workers and their representatives have the right to ask for an inspection without OSHA telling their employer who filed the complaint.* It is a violation of the OSH Act for an employer to fire, demote, transfer or retaliate in any way against a worker for filing a complaint or using other OSHA rights.

A complaint can be filed in a number of ways:

1) *Mail or submit the OSHA Complaint Form* – Download the OSHA complaint form from our website (or request a copy from your local OSHA regional or area office), complete it and then fax or mail it back to your nearest OSHA regional or area office. Written complaints that report a serious hazard and are signed by a current worker or representative and submitted to the closest OSHA area office are given priority and are more likely to result in on-site OSHA inspections. A worker or their representative can request (on the form) that OSHA not let their employer know who filed the complaint. Please include your name, address and telephone number so we can contact you to follow up. This information is confidential.
2) *Online* – Go to the online Complaint Form on the OSHA website, at www.osha.gov/pls/osha7/eComplaintForm.html. Complaints that are sent in online will most likely be investigated using OSHA's phone/fax system whereby the employer is contacted by phone or fax (not an actual inspection) about the hazard. *A written complaint that reports a serious hazard and is signed by a current worker(s) or their representative and mailed or otherwise submitted to an OSHA area or regional office is more likely to result in an on-site OSHA inspection.* Complaints received online from workers in OSHA-approved state plan states will be forwarded to the appropriate state plan for response.
3) *Telephone* – Call your local OSHA regional or area office at 1-800-321-OSHA (6742). OSHA staff can discuss your complaint and

respond to any questions you have. *If there is an emergency or the hazard is immediately life-threatening, call your local OSHA regional or area office.*

Who Else Can File a Complaint?

Employee representatives, for the purposes of filing a complaint, are defined as any of the following:

- An authorized representative of the employee bargaining unit, such as a certified or recognized labor organization.
- An attorney acting for an employee.
- Any other person acting in a bona fide representative capacity, including, but not limited to, members of the clergy, social workers, spouses and other family members, health care providers and government officials or nonprofit groups and organizations acting upon specific complaints or injuries from individuals who are employees. In general, the affected employee should have requested, or at least approved, the filing of the complaint on his or her behalf.

In addition, anyone who knows about a workplace safety or health hazard may report unsafe conditions to OSHA, and OSHA will investigate the concerns reported.

Rights of Workers during an Inspection

During an inspection, workers or their representatives have the following rights:

- Have a representative of employees, such as the safety steward of a labor organization, go along on the inspection;
- Talk privately with the inspector; and
- Take part in meetings with the inspector before and after the inspection.

When there is no authorized employee representative, the OSHA inspector must talk confidentially with a reasonable number of workers during the inspection.

Workers are encouraged to:

- Point out hazards;
- Describe injuries or illnesses that resulted from these hazards;
- Discuss past worker complaints about hazards; and
- Inform the inspector of working conditions that are not normal during the inspection.

FOLLOWING THE INSPECTION

At the end of the inspection, the OSHA inspector will meet with the employer and the employee representatives in a closing conference to discuss any violations found and possible methods by which any hazards found will be abated. If it is not practical to hold a joint conference, the compliance officer will hold separate conferences.

When the OSHA area director determines that there has been a violation of OSHA standards, regulations, or other requirements, the area director issues a citation and notification of proposed penalty to an employer. A citation includes a description of the violation and the date by when the corrective actions must be taken. Depending on the situation, OSHA can classify a violation as serious, willful, or repeat. The employer can also be cited for failing to correct a violation for which it has already been cited. Employers must post a copy of a citation in the workplace where employees will see it.

Workers' Rights following Issuance of Citations

Workers and employers can contest citations once they are issued to the employer. Workers may only contest the amount of time the employer is given to correct the hazard. Workers or their representatives must file a notice of contest with the OSHA area office within 15 days of the issuance of a citation.

Employers have the right to challenge whether there is a violation, how the violation is classified, the amount of any penalty, what the employer must do to correct the violation and how long they have to fix it. Workers or their representatives may participate in this appeals process by electing "party status." This is done by filing a written notice with the Occupational Safety and Health Review Commission (OSHRC).

The OSHRC hears appeals of OSHA citations. They are an independent agency separate from the Department of Labor. For more information, write to:

U.S. Occupational Safety and Health
Review Commission
1120 20th Street NW, 9th Floor
Washington, DC 20036
Phone: 202-606-5400 Fax: 202-606-5050
www.oshrc.gov

Right to Information if No Inspection is Conducted or No Citation Issued

The OSHA area director evaluates complaints from employees or their representatives according to the procedures defined in the OSHA Field Operations Manual. If the area director decides not to inspect the workplace, he or she will send a letter to the complainant explaining the decision and the reasons for it.

OSHA will inform complainants that they have the right to request a review of the decision by the OSHA regional administrator. Similarly, in the event that OSHA decides not to issue a citation after an inspection, employees have a right to further clarification from the area director and an informal review by the regional administrator.

Right to Use Your Rights: Protection against Retaliation Whistleblower Protection

The OSH Act prohibits employers from retaliating against their employees for using their rights under the OSH Act. These rights include filing an OSHA complaint, participating in an inspection or talking to the inspector, seeking access to employer exposure and injury records, raising a safety or health issue with the employer, or any other workers' rights described above.

Protection from retaliation means that an employer cannot punish workers by taking "adverse action", such as:

- Firing or laying off;
- Blacklisting;
- Demoting;
- Denying overtime or promotion;
- Disciplining;
- Denying benefits;
- Failing to hire or rehire;
- Intimidation;
- Making threats;
- Reassignment affecting prospects for promotion; or
- Reducing pay or hours.

You can file a complaint alleging retaliation with OSHA if your employer has punished you for using any employee rights established under the OSH Act. *If you have been retaliated against for using your rights, you must file a complaint with OSHA within 30 calendar days from the date the retaliatory decision has been both made and communicated to you (the worker). Contact your local OSHA office by calling, within 30 days of the alleged retaliation, 1-800-321-OSHA (6742), or send a letter to your closest regional or area office. No form is required.* In states with approved state plans, employees may file a complaint with both the State and Federal OSHA.

Following a complaint, OSHA will contact the complainant and conduct an interview to determine whether an investigation is necessary.

If the evidence shows that the employee has been retaliated against for exercising safety and health rights, OSHA will ask the employer to restore that worker's job, earnings, and benefits. If the employer refuses, OSHA may take the employer to court. In such cases, a Department of Labor attorney will represent the employee to obtain this relief.

If There Is a Dangerous Situation at Work

If you believe working conditions are unsafe or unhealthful, we recommend that you bring the conditions to your employer's attention, if possible.

You may file a complaint with OSHA concerning a hazardous working condition at any time. However, you should not leave the worksite merely because you have filed a complaint. If the condition clearly presents a risk of death or serious physical harm, there is not sufficient time for OSHA to inspect, and, where possible, you have brought the condition to the attention of your employer, you may have a legal right to refuse to work in a situation in which you would be exposed to the hazard.

If a worker, with no reasonable alternative, refuses in good faith to expose himself or herself to a dangerous condition, he or she would be protected from subsequent retaliation. The condition must be of such a nature that a reasonable person would conclude that there is a real danger of death or serious harm and that there is not enough time to contact OSHA and for OSHA to inspect. Where possible, the employee must have also sought from his employer, and been unable to obtain, a correction of the condition. For more information, go to www.osha.gov/workers.html.

Additional Whistleblower Protections

Since passage of the OSH Act in 1970, Congress has expanded OSHA's whistleblower protection authority to protect workers from retaliation under 22 federal laws. These laws protect employees who report violations of various workplace safety, airline, commercial motor carrier, consumer product, environmental, financial reform, healthcare reform, nuclear, pipeline, public transportation agency, railroad, maritime and securities laws. Complaints must be reported to OSHA within set timeframes following the retaliatory action, as prescribed by each law.

These laws, and the number of days employees have to file a complaint, are:

WORKER, ENVIRONMENTAL AND NUCLEAR SAFETY LAWS

- *Asbestos Hazard Emergency Response Act* (AHERA) (90 days). Provides retaliation protection for individuals who report violations of environmental laws relating to asbestos in public or private nonprofit elementary and secondary school systems.
- *Clean Air Act* (CAA) (30 days). Provides retaliation protection for employees who, among other things, report violations of this law,

which provides for the development and enforcement of standards regarding air quality and air pollution.

- *Comprehensive Environmental Response, Compensation, and Liability Act* (CERCLA) (30 days). Protects employees who report regulatory violations involving accidents, spills, and other emergency releases of pollutants into the environment. The law also protects employees who report violations related to the cleanup of uncontrolled or abandoned hazardous waste sites.
- *Energy Reorganization Act* (ERA) (180 days). Protects certain employees in the nuclear industry who report violations of the Atomic Energy Act (AEA). Protected employees include employees of operators, contractors and subcontractors of nuclear power plants licensed by the Nuclear Regulatory Commission, and employees of contractors working with the Department of Energy under a contract pursuant to the Atomic Energy Act.
- *Federal Water Pollution Control Act* (FWPCA) (also known as the Clean Water Act) (30 days). Provides retaliation protection for employees who, among other things, report violations of the law controlling water pollution.
- *Occupational Safety and Health Act of 1970* (30 days). Provides retaliation protection for employees who exercise a variety of rights guaranteed under this law, such as filing a safety and health complaint with OSHA and participating in an inspection.
- *Safe Drinking Water Act* (SDWA) (30 days). Provides retaliation protection for employees who, among other things, report violations of this law, which requires that all drinking water systems assure that their water is potable, as determined by the Environmental Protection Agency.
- *Solid Waste Disposal Act* (SWDA) (also known as the Resource Conservation and Recovery Act) (30 days). Provides retaliation protection for employees who, among other things, report violations of the law regulating the disposal of solid waste.
- *Toxic Substances Control Act* (TSCA) (30 days). Provides retaliation protection for employees who, among other things, report violations of regulations involving the manufacture, distribution, and use of certain toxic substances.

TRANSPORTATION INDUSTRY LAWS

- *Federal Railroad Safety Act* (FRSA) (180 days). Provides protection to employees of railroad carriers and contractors and subcontractors of those carriers who report an alleged violation of any federal law, rule, or regulation relating to railroad safety or security, or gross fraud, waste, or abuse of federal grants or other public funds intended to be used for railroad safety or security; report, in good faith, a hazardous safety or security condition; refuse to violate or assist in the violation of any federal law, rule, or regulation relating to railroad safety or security; refuse to work when confronted by a hazardous safety or security condition related to the performance of the employee's duties (under imminent danger circumstances); request prompt medical or first-aid treatment for employment-related injuries; are disciplined for requesting medical or first-aid treatment or for following an order or treatment plan of a treating physician.
- *International Safe Container Act* (ISCA) (60 days). Provides retaliation protection for employees who report violations of this law, which regulates shipping containers.
- *Moving Ahead for Progress in the 21st Century Act (MAP-21)* (180 days). Prohibits retaliation by motor vehicle manufacturers, part suppliers, and dealerships against employees for providing information to the employer or the U.S. Department of Transportation about motor vehicle defects, noncompliance, or violations of the notification or reporting requirements enforced by the National Highway Traffic Safety Administration or for engaging in related protected activities as set forth in the provision.
- *National Transit Systems Security Act (NTSSA)* (180 days). Provides protection to public transit employees who, among other things, report an alleged violation of any federal law, rule, or regulation relating to public transportation agency safety or security, or fraud, waste, or abuse of federal grants or other public funds intended to be used for public transportation safety or security; refuse to violate or assist in the violation of any federal law, rule, or regulation relating to public transportation safety or security; report a hazardous safety or security condition; refuse to work when confronted by a hazardous safety or security condition related to the performance of the employee's duties (under imminent danger circumstances).

- *Pipeline Safety Improvement Act of 2002 (PSIA)* (180 days). Provides retaliation protection for employees who report violations of the federal laws regarding pipeline safety and security or who refuse to violate such provisions.
- *Seaman's Protection Act (SPA)* (180 days). Seamen are protected, among other things, for reporting to the Coast Guard or other federal agency a reasonably believed violation of a maritime safety law or regulation prescribed under that law or regulation. The law also protects work refusals where the employee reasonably believes an assigned task would result in serious injury or impairment of health to the seaman, other seamen, or the public and when the seaman sought, and was unable to obtain correction of the unsafe conditions.
- Surface Transportation Assistance Act (STAA) (180 days). Provides retaliation protection for truck drivers and other employees relating to the safety of commercial motor vehicles. Coverage includes all buses for hire and freight trucks with a gross vehicle weight greater than 10,001 pounds.
- Wendell H. Ford Aviation Investment and Reform Act for the 21st Century (AIR21) (90 days). Provides retaliation protection for employees of air carriers, contractors, or subcontractors of air carriers who, among other things, raise safety concerns.

FRAUD PREVENTION LAWS

- *Affordable Care Act (ACA)* (180 days). Protects employees who report violations of any provision of Title I of the ACA, including but not limited to retaliation based on an individual's receipt of health insurance subsidies, the denial of coverage based on a preexisting condition, or an insurer's failure to rebate a portion of an excess premium.
- *Consumer Financial Protection Act of 2010 (CFPA), Section 1057 of the Dodd-Frank Wall Street Reform and Consumer Protection Act* (180 days). Protects employees who report perceived violations of any provision of the *Dodd-Frank Act*, which encompasses nearly every aspect of the financial services industry. The law also protects employees who report violations of any rule, order, standard or

prohibition prescribed by the Bureau of Consumer Financial Protection.

- *Section 806 of the Sarbanes-Oxley Act of 2002 (SOX)* (180 days). Protects employees of certain companies who report alleged mail, wire, bank or securities fraud; violations of the Securities and Exchange Commission (SEC) rules and regulations; or violations of Federal laws related to fraud against shareholders. The law covers employees of publicly traded companies and companies required to file certain reports with the SEC.

Consumer Safety Laws

- *Consumer Product Safety Improvement Act* (CPSIA) (180 days). Protects employees who report to their employer, the federal government, or a state attorney general reasonably perceived violations of any statute or regulation within the jurisdiction of the Consumer Product Safety Commission (CPSC). CPSIA covers employees of consumer product manufacturers, importers, distributors, retailers, and private labelers.
- *FDA Food Safety Modernization Act* (FSMA) (180 days). Protects employees of food manufacturers, distributors, packers, and transporters for reporting a violation of the Food, Drug, and Cosmetic Act, or a regulation promulgated under this law. Employees are also protected from retaliation for refusing to participate in a practice that violates this law.

If you believe that you have been retaliated against, call 1-800-321-OSHA (6742) to be connected to the nearest OSHA office to report your complaint. For more information, visit OSHA's Whistleblower page at www.whistle blowers.gov.

OSHA Assistance, Services and Programs

OSHA has a great deal of information to assist employers in complying with their responsibilities under OSHA law. Several OSHA programs and services can help employers identify and correct job hazards, as well as improve their injury and illness prevention program.

Establishing an Injury and Illness Prevention Program

The key to a safe and healthful work environment is a comprehensive injury and illness prevention program.

Injury and illness prevention programs are systems that can substantially reduce the number and severity of workplace injuries and illnesses, while reducing costs to employers. Thousands of employers across the United States already manage safety using injury and illness prevention programs, and OSHA believes that all employers can and should do the same. Thirty-four states have requirements or voluntary guidelines for workplace injury and illness prevention programs. Most successful injury and illness prevention programs are based on a common set of key elements. These include management leadership, worker participation, hazard identification, hazard prevention and control, education and training, and program evaluation and improvement. Visit OSHA's Injury and Illness Prevention Programs web page at www.osha.gov/dsg/topics/safetyhealth for more information.

Compliance Assistance Specialists

OSHA has compliance assistance specialists throughout the nation located in most OSHA offices. Compliance assistance specialists can provide information to employers and workers about OSHA standards, short educational programs on specific hazards or OSHA rights and responsibilities, and information on additional compliance assistance resources. For more details, visit www.osha.gov/ dcsp/compliance_assistance/cas.html or call 1-800- 321-OSHA (6742) to contact your local OSHA office.

Free On-site Safety and Health Consultation Services for Small Business

OSHA's On-site Consultation Program offers free and confidential advice to small and medium-sized businesses in all states across the country, with priority given to high-hazard worksites. Each year, responding to requests from small employers looking to create or improve their safety and health management programs, OSHA's On-site Consultation Program conducts over 29,000 visits to small business worksites covering over 1.5 million workers across the nation.

On-site consultation services are separate from enforcement and do not result in penalties or citations. Consultants from state agencies or universities work with employers to identify workplace hazards, provide advice on compliance with OSHA standards, and assist in establishing safety and health management programs.

For more information, to find the local On-site Consultation office in your state, or to request a brochure on consultation services, visit www. osha.gov/ consultation, or call 1-800-321-OSHA (6742).

Under the consultation program, certain exemplary employers may request participation in OSHA's *Safety and Health Achievement Recognition Program (SHARP)*. Eligibility for participation includes, but is not limited to, receiving a full-service, comprehensive consultation visit, correcting all identified hazards and developing an effective safety and health management program. Worksites that receive SHARP recognition are exempt from programmed inspections during the period that the SHARP certification is valid.

Cooperative Programs

OSHA offers cooperative programs under which businesses, labor groups and other organizations can work cooperatively with OSHA. To find out more about any of the following programs, visit www.osha.gov/ dcsp/compliance_ assistance/index_programs.html.

Strategic Partnerships and Alliances

The OSHA Strategic Partnerships (OSP) provide the opportunity for OSHA to partner with employers, workers, professional or trade associations, labor organizations, and/or other interested stakeholders. OSHA Partnerships are formalized through unique agreements designed to encourage, assist, and recognize partner efforts to eliminate serious hazards and achieve model workplace safety and health practices. Through the Alliance Program, OSHA works with groups committed to worker safety and health to prevent workplace fatalities, injuries and illnesses by developing compliance assistance tools and resources to share with workers and employers, and educate workers and employers about their rights and responsibilities.

Voluntary Protection Programs (VPP)

The VPP recognize employers and workers in private industry and federal agencies who have implemented effective safety and health management programs and maintain injury and illness rates below the national average for their respective industries. In VPP, management, labor, and OSHA work cooperatively and proactively to prevent fatalities, injuries, and illnesses through a system focused on: hazard prevention and control, worksite analysis, training, and management commitment and worker involvement.

Occupational Safety and Health Training

The OSHA Training Institute in Arlington Heights, Illinois, provides basic and advanced training and education in safety and health for federal and state compliance officers, state consultants, other federal agency personnel and private sector employers, workers, and their representatives. In addition, 27 OSHA Training Institute Education Centers at 42 locations throughout the United States deliver courses on OSHA standards and occupational safety and health issues to thousands of students a year.

For more information on training, contact the OSHA Directorate of Training and Education, 2020 Arlington Heights Road, Arlington Heights, IL 60005; call 1-847- 297-4810; or visit www.osha.gov/otiec.

OSHA Educational Materials

OSHA has many types of educational materials in English, Spanish, Vietnamese and other languages available in print or online. These include:

- Brochures/booklets that cover a wide variety of job hazards and other topics;
- Fact Sheets, which contain basic background information on safety and health hazards;

- Guidance documents that provide detailed examinations of specific safety and health issues;
- Online Safety and Health Topics pages;
- Posters;
- Small, laminated QuickCardsTM that provide brief safety and health information; and
- *QuickTakes*, OSHA's free, twice-monthly online newsletter with the latest news about OSHA initiatives and products to assist employers and workers in finding and preventing workplace hazards. To sign up for *QuickTakes* visit www.osha.gov/quicktakes.

To view materials available online or for a listing of free publications, visit www.osha.gov/publications. You can also call 1-800-321-OSHA (6742) to order publications.

OSHA's website also has a variety of eTools. These include utilities such as expert advisors, electronic compliance assistance, videos and other information for employers and workers. To learn more about OSHA's safety and health tools online, visit www.osha.gov.

NIOSH Health Hazard Evaluation Program

Getting Help with Health Hazards

The National Institute for Occupational Safety and Health (NIOSH) is a federal agency that conducts scientific and medical research on workers' safety and health. At no cost to employers or workers, NIOSH can help identify health hazards and recommend ways to reduce or eliminate those hazards in the workplace through its Health Hazard Evaluation (HHE) Program.

Workers, union representatives and employers can request a NIOSH HHE. An HHE is often requested when there is a higher than expected rate of a disease or injury in a group of workers. These situations may be the result of an unknown cause, a new hazard, or a mixture of sources. To request a NIOSH Health Hazard Evaluation go to www.cdc.gov/ niosh/hhe/request.html. To find out more about the Health Hazard Evaluation Program:

- Call (513) 841-4382, or to talk to a staff member in Spanish, call (513) 841-4439; or

- Send an email to HHERequestHelp@cdc.gov.

How to Contact OSHA

For questions or to get information or advice, to report an emergency, report a fatality or catastrophe, order publications, sign up for OSHA's e-newsletter QuickTakes, or to file a confidential complaint, contact your nearest OSHA office, visit www.osha.gov or call OSHA at 1-800-321-OSHA (6742), TTY 1-877-889-5627.

For assistance, contact us. We are OSHA. We can help. It's confidential.

OSHA Regional Offices

Region I
Boston Regional Office
(CT*, ME, MA, NH, RI, VT*)
JFK Federal Building, Room E340
Boston, MA 02203
(617) 565-9860 (617) 565-9827 Fax

Region II
New York Regional Office
(NJ*, NY*, PR*, VI*)
201 Varick Street, Room 670
New York, NY 10014
(212) 337-2378 (212) 337-2371 Fax

Region III
Philadelphia Regional Office
(DE, DC, MD*, PA, VA*, WV)
The Curtis Center
170 S. Independence Mall West
Suite 740 West
Philadelphia, PA 19106-3309
(215) 861-4900 (215) 861-4904 Fax

Region IV
Atlanta Regional Office
(AL, FL, GA, KY*, MS, NC*, SC*, TN*) 61 Forsyth Street, SW, Room 6T50 Atlanta, GA 30303
(678) 237-0400 (678) 237-0447 Fax

Region V
Chicago Regional Office
(IL*, IN*, MI*, MN*, OH, WI)
230 South Dearborn Street
Room 3244
Chicago, IL 60604
(312) 353-2220 (312) 353-7774 Fax

Region VI
Dallas Regional Office
(AR, LA, NM*, OK, TX)
525 Griffin Street, Room 602
Dallas, TX 75202
(972) 850-4145 (972) 850-4149 Fax
(972) 850-4150 FSO Fax

Region VII
Kansas City Regional Office (IA*, KS, MO, NE)
Two Pershing Square Building 2300 Main Street, Suite 1010 Kansas City, MO 64108-2416 (816) 283-8745 (816) 283-0547 Fax

Region VIII
Denver Regional Office
(CO, MT, ND, SD, UT*, WY*) Cesar Chavez Memorial Building 1244 Speer Boulevard, Suite 551 Denver, CO 80204
(720) 264-6550 (720) 264-6585 Fax

Region IX
San Francisco Regional Office
(AZ*, CA*, HI*, NV*, and American Samoa,
Guam and the Northern Mariana Islands)
90 7th Street, Suite 18100
San Francisco, CA 94103

(415) 625-2547 (415) 625-2534 Fax
Region X
Seattle Regional Office
(AK*, ID, OR*, WA*)
300 Fifth Avenue, Suite 1280
Seattle, WA 98104
(206) 757-6700 (206) 757-6705 Fax

* These states and territories operate their own OSHA-approved job safety and health plans and cover state and local government employees as well as private sector employees. The Connecticut, Illinois, New Jersey, New York and Virgin Islands programs cover public employees only. (Private sector workers in these states are covered by Federal OSHA). States with approved programs must have standards that are identical to, or at least as effective as, the Federal OSHA standards.

Note: To get contact information for OSHA area offices, OSHA-approved state plans and OSHA consultation projects, please visit us online at www.osha.gov or call us at 1-800-321-OSHA (6742).

In: The Occupational Safety and Health Act
Editor: Christy Knight
ISBN: 978-1-63482-637-2

Chapter 5

EMPLOYER RIGHTS AND RESPONSIBILITIES FOLLOWING A FEDERAL OSHA INSPECTION*

Occupational Safety and Health Administration

This publication provides a general overview of employer rights and responsibilities following a Federal OSHA inspection. It does not alter or determine compliance responsibilities in OSHA standards or the *Occupational Safety and Health Act of 1970*. Because interpretations and enforcement policy may change over time, you should consult current OSHA administrative interpretations and decisions by the Occupational Safety and Health Review Commission and the courts for additional guidance on OSHA compliance requirements.

AFTER A FEDERAL OSHA INSPECTION

This pamphlet contains important information regarding employer rights and responsibilities following a Federal Occupational Safety and Health Administration (OSHA) inspection under the *Occupational Safety and Health*

---

* This is an edited, reformatted and augmented version of a publication, OSHA 3000-09R, issued by the U.S. Department of Labor, 2015.

Act of 1970 (OSH Act), as amended. Under the OSH Act, employers have the responsibility to provide a safe workplace.

An OSHA compliance safety and health officer (CSHO) conducts an inspection of your workplace, in accordance with the OSH Act. After the inspection, the CSHO reports the findings to the OSHA area director who evaluates them. If a violation exists, OSHA will issue you a *Citation and Notification of Penalty* detailing the exact nature of the violation(s) and any associated penalties. A citation informs you of the alleged violation, sets a proposed time period within which to correct the violation, and proposes the appropriate dollar penalties.

The information in this booklet can and should be used as a discussion guide during your closing conference with the CSHO. For each apparent violation found during the inspection, the compliance officer has discussed or will discuss the following with you:

- Nature of the violation;
- Possible abatement measures you may take to correct the violative condition;
- Possible abatement dates you may be required to meet; and
- Any penalties that the area director may issue.

The CSHO is a highly trained professional who can help you recognize and evaluate hazards as well as suggest appropriate methods of correcting violations. To minimize worker exposure to possible hazardous conditions, abatement efforts should always begin as soon as possible.

Posting Requirements

When you receive a Citation and Notification of Penalty, you must post the citation (or a copy of it) at or near the place where each violation occurred to make employees aware of the hazards to which they may be exposed. The citation must remain posted in a place where employees can see it, for three working days or until the violation is corrected, whichever is longer. (Saturdays, Sundays, and Federal holidays are not counted as working days.) You must comply with these posting requirements even if you contest the citation.

The abatement certification documents – such as abatement certifications, abatement plans and progress reports – also must be posted at or near the place where the violation occurred. For moveable equipment found to be in violation and where the posting of violations would be difficult or impractical, the employer has the option to identify the equipment with a "Warning" tag specified in the abatement verification regulation, Title 29 *Code of Federal Regulations* (CFR) 1903.19(i).

EMPLOYER OPTIONS

As an employer who has been cited, you may take either of the following courses of action:

- If you agree to the Citation and Notification of Penalty, you must correct the condition by the date set in the citation and pay the penalty, if one is proposed.
- If you do not agree, you have 15 working days from the date you receive the citation to contest in writing any or all of the following:
  - Citation;
  - Proposed penalty; and/or
  - Abatement date.

OSHA encourages you to call the local OSHA area office upon receipt of the citations to discuss any concerns. Before deciding to contest the citation, *you may request an informal conference with the OSHA area director within the 15-working-day period to discuss any issues related to the Citation and Notification of Penalty*. (See the following section on Informal Conference and Settlement).

OSHA will inform the affected employee representatives of the informal conference or contest.

INFORMAL CONFERENCE AND SETTLEMENT

Before deciding whether to file a Notice of Intent to Contest, you may request an informal conference with the OSHA area director to discuss the

Citation and Notification of Penalty. You may use this opportunity to do any of the following:

- Obtain a better explanation of the violations cited;
- Obtain a more complete understanding of the specific standards that apply;
- Negotiate and enter into an informal settlement agreement;
- Discuss ways to correct violations;
- Discuss issues concerning proposed penalties;
- Discuss proposed abatement dates;
- Resolve disputed citations and penalties, (thereby eliminating the need for the more formal procedures associated with litigation before the Occupational Safety and Health Review Commission); and
- Obtain answers to any other questions you may have.

OSHA encourages you to take advantage of the opportunity to have an informal conference if you foresee any difficulties in complying with any part of the citation. *Please note, however, that an informal conference must be held within the 15-working-day Notice of Intent to Contest period and will neither extend the 15-workingday contest period nor take the place of the filing of a written notice if you desire to contest.* Employee representative(s) have the right to participate in any informal conference or negotiations between the regional administrator or area director and the employer.

If you agree that the cited violations exist, but you have a valid reason for wishing to extend the abatement date(s), you may discuss this with the area director in an informal conference. He or she may issue an amended citation that changes the abatement date prior to the expiration of the 15-working-day period without your filing a Notice of Intent to Contest.

If you do not contest within 15 working days, your citation will become a final order not subject to review by any court or agency. After this occurs, the OSHA area director may continue to provide you with information and assistance on how to abate the hazards cited in your citation, but may not amend or change any citation or penalty which has become a final order. The area director may only advise you on abatement methods or extend the time you need to abate the violation. (*See Petition for Modification of Abatement*).

Whenever the employer, an affected employee, or employee representative requests an informal conference, *all* the parties shall be afforded the opportunity to participate fully. If either party chooses not to participate in the informal conference, that party forfeits the right to be consulted before decisions are made that affect the citations. If the requesting party objects to the attendance of the other party, OSHA may hold separate informal conferences. During a joint informal conference, separate or private discussions will be permitted if either party requests them. Informal conferences may be held by any means practical.

TYPES OF VIOLATIONS

The following general information defines the types of violations and explains the actions you may take if you receive a citation as the result of an inspection.

In settling a penalty, OSHA has a policy of reducing penalties for small employers and those acting in good faith.

Willful: A willful violation exists under the OSH Act where an employer has demonstrated either an intentional disregard for the requirements of the OSH Act or a plain indifference to employee safety and health. Penalties range from $5,000 to $70,000 per willful violation.

Serious: Section 17(k) of the OSH Act provides that "a serious violation shall be deemed to exist in a place of employment if there is a substantial probability that death or serious physical harm could result from a condition which exists, or from one or more practices, means, methods, operations, or processes which have been adopted or are in use, in such place of employment unless the employer did not, and could not with the exercise of reasonable diligence, know of the presence of the violation." OSHA may propose a penalty of up to $7,000 for each violation.

Other-Than-Serious: This type of violation is cited in situations where the accident/incident or illness that would be most likely to result from a hazardous condition would probably not cause death or serious physical harm, but would have a direct and immediate relationship to the safety and health of employees. OSHA may impose a penalty of up to $7,000 for each violation.

De Minimis: De minimis conditions are those where an employer has implemented a measure different from one specified in a standard, that has no direct or immediate relationship to safety or health. These conditions do not result in citations or penalties.

Failure to Abate: A failure to abate violation exists when a previously cited hazardous condition, practice or non-complying equipment has not been brought into compliance since the prior inspection (i.e., the violation remains continuously uncorrected) and is discovered at a later inspection. If, however, the violation was corrected, but later reoccurs, the subsequent occurrence is a repeated violation. OSHA may impose a penalty of up to $7,000 per day for each violation.

Repeated: An employer may be cited for a repeated violation if that employer has been cited previously, within the last five years, for the same or a substantially similar condition or hazard and the citation has become a final order of the Occupational Safety and Health Review Commission (OSHRC). A citation may become a final order by operation of law when an employer does not contest the citation, or pursuant to court decision or settlement. Repeated violations can bring a civil penalty of up to $70,000 for each violation.

How to Comply

For violations you do not contest, you must:

- promptly notify the OSHA area director by letter, signed by a member of management, that you have taken the appropriate corrective action within the time set forth in the citation, and
- pay any penalties itemized.

The notification you send the area director is referred to as *Abatement Certification.* For OtherThan-Serious violations, this may be a signed letter identifying the inspection number and the citation item number and noting that you corrected the violation by the date specified on the citation. For more serious violations (such as Serious, Willful, Repeated, or Failure to Abate), abatement certification requires more detailed proof.

If the employer has abatement questions after the inspection, the area director must ensure that additional information, if available, is provided to the employer as soon as possible.

Employers also can find guidance on abatement verification on OSHA's website at www.osha.gov/ Publications/Abate/abate.html.

When the citation permits an extended time for abatement, you must ensure that employees are adequately protected during this time. For example, the citation may require the immediate use of personal protective equipment by employees while engineering controls are being installed. When indicated on the citation, you must also provide OSHA with an abatement plan (steps you will take to protect employees and correct the hazards) and periodic progress reports on your actions.

The penalties itemized on the Citation and Notification of Penalty are payable within 15 working days of receipt of the penalty notice. If, however, you contest the citation or penalty in good faith, OSHA will suspend abatement and payment of penalties for those items contested until the Occupational Safety and Health Review Commission (OSHRC), or a higher court, issues a final order or decision. The OSHRC is an independent agency and is *not* a part of the U.S. Department of Labor. The final order of OSHRC will either uphold, modify, or eliminate the citations and/or penalties. Penalties for items not contested, however, are still due within 15 working days. (For further details, see the following section on *How to Contest Citations*.)

Payment should be made by check or money order payable to DOL-OSHA. Please indicate on your payment the OSHA number from the upper right-hand corner of your citation and *send it to the OSHA area office listed on the Citation and Notification of Penalty.*

How to Contest Citations

If you wish to contest any portion of your citation, you must submit a Notice of Intent to Contest in writing to the OSHA area office within 15 working days after receipt of the Citation and Notification of Penalty. This applies even if you have stated your disagreement with a citation, penalty, or abatement date during a telephone conversation or an informal conference.

The Notice of Intent to Contest must clearly state what is being contested – the citation, the penalty, the abatement date, or any combination of these factors. In addition, the notice must state whether all the violations on the citation, or just specific violations, are being contested. (For example, “I wish to contest the citation and penalty proposed for items 3 and 4 of the citation issued June 27, 2011”).

Your contest must be made in good faith. OSHA will not consider a contest filed solely to avoid your responsibilities for abatement or payment of penalties.

A proper contest of any item suspends your legal obligation to abate and pay until the item contested has been resolved. If you contest only the dates indicated on the citation or if you contest only some items on the citation, you must correct the other items by the abatement date and pay the corresponding penalties within 15 days of notification.

After you file a Notice of Intent to Contest, your case is officially in litigation. If you wish to settle the case, you may contact the OSHA area director who will give you the name of the attorney handling your case for OSHA. All settlements of contested cases are negotiated between you and the attorney according to the rules of procedure of the OSHRC.

THE CONTEST PROCESS

If you file the written Notice of Intent to Contest within the required 15 working days, the OSHA area director forwards your case to the OSHRC. The OSHRC hears employer contests of OSHA citations. They are an independent agency separate from the Department of Labor. The OSHRC assigns the case to an administrative law judge who usually will schedule a hearing in a public place close to your workplace. Both employers and employees have the right to participate in this hearing, which contains all the elements of a trial, including examination and cross-examination of witnesses. You may choose to represent yourself or have an attorney represent you. The administrative law judge may affirm, modify, or eliminate any contested items of the citation or penalty.

As with any other legal procedure, there is an appeals process. Once the administrative law judge has ruled, any party to the case may request a further review by the full OSHRC. In addition, any of the three commissioners may, on his or her own motion, bring the case before the entire OSHRC for review. The OSHRC's ruling, in turn, may be appealed to the Federal circuit court in which the case arose or for the Federal circuit where the employer has his or her principal office.

For more information, write to:

U.S. Occupational Safety and Health
Review Commission
1120 20th Street, NW, 9th Floor
Washington, DC 20036
Phone: 202-606-5400 Fax: 202-606-5050
www.oshrc.gov

Petition for Modification of Abatement

OSHA assigns abatement dates on the basis of the best information available when issuing the citation. If you are unable to meet an abatement date because of uncontrollable events or other circumstances, and the 15-working-day contest period has expired, you may file a *Petition for Modification of Abatement* (PMA) with the OSHA area director.

The petition must be in writing and must be submitted as soon as possible, *but no later than one working day after the abatement date*. To show clearly that you have made a good-faith effort to comply, the PMA must include all of the following information before OSHA considers it:

- Steps you have taken to achieve compliance, and dates they were taken;
- Additional time you need to comply; n Why you need additional time;
- Interim steps you are taking to safeguard your workers against the cited hazard(s) until the abatement; and
- A certification that the petition has been posted, the date of posting and, when appropriate, a statement that the petition has been furnished to an authorized representative of the affected employees. The petition must remain posted for 10 working days, during which employees may file an objection.

The OSHA area director may grant or oppose a PMA. If it is opposed, it automatically becomes a contested case before the OSHRC. If a PMA is granted, OSHA may conduct a monitoring inspection to ensure that conditions are as they have been described and that adequate progress has been made toward abatement. The OSHA area office may provide additional information on PMAs.

What Workers Can Do

Employees or their authorized representatives may contest any or all of the abatement dates set for violations if they believe them to be unreasonable. A written Notice of Intent to Contest must be filed with the OSHA area director within 15 working days after the employer receives the citation.

The Filing of an Employee Contest Does Not Suspend the Employer's Obligation to Abate

Employees also have the right to object to a PMA. Such objections must be in writing and must be sent to the area office within 10 days of service or posting. OSHA will not make a decision regarding the PMA until the Review Commission resolves the issue.

FOLLOW-UP INSPECTIONS AND FAILURE TO ABATE

If you receive a citation, a follow-up inspection may be conducted to verify that you have done the following:

- Posted the citation as required,
- Corrected the violations as required in the citation, and/or
- Protected employees adequately and made appropriate progress in correcting hazards during multistep or lengthy abatement periods.

In addition to providing for penalties for failure-topost citations and failure-to-abate violations, the OSH Act clearly states that you have a *continuing responsibility* to comply with the OSH Act and assure your employees safe and healthful working conditions. OSHA will cite any new violations discovered during a follow-up inspection.

EMPLOYER RESPONSIBILITIES

Employers have the responsibility to provide a safe workplace. Employers MUST provide their employees with a workplace that does not have serious hazards and must follow all OSHA safety and health standards. Employers must find and correct safety and health problems.

Employers MUST also:

- Prominently display the official OSHA *Job Safety and Health – It's the Law* poster that describes rights and responsibilities under the OSH Act. *This poster is free and can be downloaded from www.osha.gov.*

- Inform workers about chemical hazards through training, labels, alarms, color-coded systems, chemical information sheets and other methods.
- Provide safety training to workers in a language and vocabulary they can understand.
- Keep accurate records of work-related injuries and illnesses.
- Perform tests in the workplace, such as air sampling, required by some OSHA standards.
- Provide required personal protective equipment at no cost to workers.*
- Provide hearing exams or other medical tests required by OSHA standards.
- Post OSHA citations and injury and illness data where workers can see them.
- Notify OSHA within 8 hours of a workplace fatality or within 24 hours of any work-related inpatient hospitalization, amputation or loss of an eye (1-800-321-OSHA [6742]).
- Not retaliate against workers for using their rights under the law, including their right to report a work-related injury or illness.

* Employers must pay for most types of required personal protective equipment.

WORKER RETALIATION

Section 11(c) of the OSH Act prohibits employers from discharging or otherwise retaliating against an employee who has exercised any right under this law, including the right to make safety and health complaints or to request an OSHA inspection. OSHA will investigate complaints from employees who believe they have been retaliated against. If the investigation discloses probable violations of worker rights, court action may follow.

Employees who believe they have been retaliated against must file their complaints with OSHA within 30 calendar days from the date the retaliatory decision has been both made and communicated to the worker. No form is needed, but workers must call OSHA within 30 days of the alleged retaliation (at 1-800-321-OSHA [6742]). In states with OSHA-approved state programs, an employee who believes he/she has been retaliated against under Section

11(c) of the OSH Act is entitled to file a complaint under both state and federal procedures.

PROVIDING FALSE INFORMATION

All information employers report to OSHA must be accurate and truthful. Providing false information on efforts to abate cited conditions or in required records is punishable under the OSH Act.

OSHA ASSISTANCE, SERVICES AND PROGRAMS

OSHA has a great deal of information to assist employers in complying with their responsibilities under OSHA law. Several OSHA programs and services can help employers identify and correct job hazards, as well as improve their injury and illness prevention program.

Establishing an Injury and Illness Prevention Program

The key to a safe and healthful work environment is a comprehensive injury and illness prevention program.

Injury and illness prevention programs are systems that can substantially reduce the number and severity of workplace injuries and illnesses, while reducing costs to employers. Thousands of employers across the United States already manage safety using injury and illness prevention programs, and OSHA believes that all employers can and should do the same. Thirty-four states have requirements or voluntary guidelines for workplace injury and illness prevention programs. Most successful injury and illness prevention programs are based on a common set of key elements. These include management leadership, worker participation, hazard identification, hazard prevention and control, education and training, and program evaluation and improvement. Visit OSHA's Injury and Illness Prevention Programs web page at www.osha.gov/dsg/topics/ safetyhealth for more information.

Compliance Assistance Specialists

OSHA has compliance assistance specialists throughout the nation located in most OSHA offices. Compliance assistance specialists can provide information to employers and workers about OSHA standards, short educational programs on specific hazards or OSHA rights and responsibilities, and information on additional compliance assistance resources. For more details, visit www.osha.gov/dcsp/compliance_assistance/cas.html or call 1-800-321-OSHA (6742) to contact your local OSHA office.

Free On-site Safety and Health Consultation Services for Small Business

OSHA's On-site Consultation Program offers free and confidential advice to small and medium-sized businesses in all states across the country, with priority given to high-hazard worksites. Each year, responding to requests from small employers looking to create or improve their safety and health management programs, OSHA's On-site Consultation Program conducts over 29,000 visits to small business worksites covering over 1.5 million workers across the nation.

On-site consultation services are separate from enforcement and do not result in penalties or citations. Consultants from state agencies or universities work with employers to identify workplace hazards, provide advice on compliance with OSHA standards, and assist in establishing safety and health management programs.

For more information, to find the local On-site Consultation office in your state, or to request a brochure on Consultation Services, visit www.osha.gov/consultation, or call 1-800-321-OSHA (6742).

Under the consultation program, certain exemplary employers may request participation in OSHA's *Safety and Health Achievement Recognition Program (SHARP)*. Eligibility for participation includes, but is not limited to, receiving a full-service, comprehensive consultation visit, correcting all identified hazards and developing an effective safety and health management program. Worksites that receive SHARP recognition are exempt from programmed inspections during the period that the SHARP certification is valid.

Cooperative Programs

OSHA offers cooperative programs under which businesses, labor groups and other organizations can work cooperatively with OSHA. To find out more about any of the following programs, visit www.osha.gov/cooperativeprograms.

Strategic Partnerships and Alliances

The OSHA Strategic Partnerships (OSP) provide the opportunity for OSHA to partner with employers, workers, professional or trade associations, labor organizations, and/or other interested stakeholders. OSHA Partnerships are formalized through unique agreements designed to encourage, assist, and recognize partner efforts to eliminate serious hazards and achieve model workplace safety and health practices. Through the Alliance Program, OSHA works with groups committed to worker safety and health to prevent workplace fatalities, injuries and illnesses by developing compliance assistance tools and resources to share with workers and employers, and educate workers and employers about their rights and responsibilities.

Voluntary Protection Programs (VPP)

The VPP recognize employers and workers in private industry and federal agencies who have implemented effective safety and health management programs and maintain injury and illness rates below the national average for their respective industries. In VPP, management, labor, and OSHA work cooperatively and proactively to prevent fatalities, injuries, and illnesses through a system focused on: hazard prevention and control, worksite analysis, training, and management commitment and worker involvement.

Occupational Safety and Health Training

The OSHA Training Institute partners with 27 OSHA Training Institute Education Centers at 42 locations throughout the United States to deliver courses on OSHA standards and occupational safety and health topics to thousands of students a year. For more information on training courses, visit www.osha.gov/otiec.

OSHA Educational Materials

OSHA has many types of educational materials in English, Spanish, Vietnamese and other languages available in print or online. These include:

- Brochures/booklets;
- Fact Sheets;
- Guidance documents that provide detailed examinations of specific safety and health issues;
- Online Safety and Health Topics pages;
- Posters;
- Small, laminated QuickCardsTM that provide brief safety and health information; and
- QuickTakes, OSHA's free, twice-monthly online newsletter with the latest news about OSHA initiatives and products to assist employers and workers in finding and preventing workplace hazards. To sign up for QuickTakes visit www.osha.gov/quicktakes.

To view materials available online or for a listing of free publications, visit www.osha.gov/ publications. You can also call 1-800-321-OSHA (6742) to order publications.

Select OSHA publications are available in e-Book format. OSHA e-Books are designed to increase readability on smartphones, tablets and other mobile devices. For access, go to www.osha.gov/ebooks.

OSHA's web site also has information on job hazards and injury and illness prevention for employers and workers. To learn more about OSHA's safety and health resources online, visit www.osha.gov or www.osha.gov/html/ a-z-index.html.

NIOSH HEALTH HAZARD EVALUATION PROGRAM

Getting Help with Health Hazards

The National Institute for Occupational Safety and Health (NIOSH) is a federal agency that conducts scientific and medical research on workers' safety and health. At no cost to employers or workers, NIOSH can help identify health hazards and recommend ways to reduce or eliminate those hazards in the workplace through its Health Hazard Evaluation (HHE) Program.

Workers, union representatives and employers can request a NIOSH HHE. An HHE is often requested when there is a higher than expected rate of a disease or injury in a group of workers. These situations may be the result of an unknown cause, a new hazard, or a mixture of sources. To request a NIOSH Health Hazard Evaluation go to www.cdc.gov/niosh/hhe/request.html. To find out more, in English or Spanish, about the Health Hazard Evaluation Program:

E-mail HHERequestHelp@cdc.gov or call 800-CDC-INFO (800-232-4636).

How to Contact OSHA

For questions or to get information or advice, to report an emergency, fatality, inpatient hospitalization, amputation, or loss of an eye, or to file a confidential complaint, contact your nearest OSHA office, visit www.osha.gov or call OSHA at 1-800-321-OSHA (6742), TTY 1-877-889-5627.

For assistance, contact us. We are OSHA. We can help.

OSHA Regional Offices

Region I
Boston Regional Office
(CT*, ME*, MA, NH, RI, VT*)
JFK Federal Building, Room E340
Boston, MA 02203
(617) 565-9860 (617) 565-9827 Fax

Region II
New York Regional Office
(NJ*, NY*, PR*, VI*)
201 Varick Street, Room 670
New York, NY 10014
(212) 337-2378 (212) 337-2371 Fax

Region III
Philadelphia Regional Office
(DE, DC, MD*, PA, VA*, WV)
The Curtis Center
170 S. Independence Mall West
Suite 740 West
Philadelphia, PA 19106-3309
(215) 861-4900 (215) 861-4904 Fax

Region IV
Atlanta Regional Office
(AL, FL, GA, KY*, MS, NC*, SC*, TN*) 61 Forsyth Street, SW, Room 6T50
Atlanta, GA 30303
(678) 237-0400 (678) 237-0447 Fax

Region V
Chicago Regional Office
(IL*, IN*, MI*, MN*, OH, WI)
230 South Dearborn Street
Room 3244
Chicago, IL 60604
(312) 353-2220 (312) 353-7774 Fax

Region VI
Dallas Regional Office
(AR, LA, NM*, OK, TX)
525 Griffin Street, Room 602
Dallas, TX 75202
(972) 850-4145 (972) 850-4149 Fax
(972) 850-4150 FSO Fax

Region VII
Kansas City Regional Office
(IA*, KS, MO, NE)
Two Pershing Square Building
2300 Main Street, Suite 1010
Kansas City, MO 64108-2416
(816) 283-8745 (816) 283-0547 Fax

Region VIII
Denver Regional Office
(CO, MT, ND, SD, UT*, WY*) Cesar Chavez Memorial Building 1244 Speer Boulevard, Suite 551 Denver, CO 80204
(720) 264-6550 (720) 264-6585 Fax

Region IX
San Francisco Regional Office
(AZ*, CA*, HI*, NV*, and American Samoa,
Guam and the Northern Mariana Islands)
90 7th Street, Suite 18100
San Francisco, CA 94103
(415) 625-2547 (415) 625-2534 Fax

Region X
Seattle Regional Office
(AK*, ID, OR*, WA*)
300 Fifth Avenue, Suite 1280
Seattle, WA 98104
(206) 757-6700 (206) 757-6705 Fax

*These states and territories operate their own OSHA-approved job safety and health plans and cover state and local government employees as well as private sector employees. The Connecticut, Illinois, Maine, New Jersey, New York and Virgin Islands programs cover public employees only. (Private sector workers in these states are covered by Federal OSHA). States with approved programs must have standards that are identical to, or at least as effective as, the Federal OSHA standards.

Note: To get contact information for OSHA area offices, OSHA-approved state plans and OSHA consultation projects, please visit us online at www.osha.gov or call us at 1-800-321-OSHA (6742).

Appendix: The Small Business Regulatory Enforcement Fairness Act of 1996 (SBREFA)

The *Small Business Regulatory Enforcement Fairness Act of 1996* (SBREFA) gives small businesses help in understanding and complying with

OSHA regulations and allows them a voice in developing new regulations. Under SBREFA, OSHA must:

- Produce Small Entity Compliance Guides for some agency rules;
- Be responsive to small business inquiries about complying with the Agency's regulations;
- Have a penalty reduction policy for small businesses;
- Involve small businesses in developing proposed rules expected to significantly affect a large number of small entities through Small Business Advocacy Review Panels; and
- Give small businesses the opportunity to challenge in court agency rules or regulations that they believe will adversely affect them.

Commenting on Enforcement Actions

Under a law passed by Congress in 1996, the Small Business Administration (SBA) has established an SBA Ombudsman and SBA Regional Fairness Boards to investigate small business complaints about Federal agency enforcement actions.

If you are a small business and believe that you have been treated unfairly by OSHA, you may file an electronic comment/complaint with the SBA Ombudsman over the Internet at: www.sba.gov/ombudsman/comments/commentform1. html. Or you may contact the SBA's Office of the National Ombudsman by:

- Toll-free Phone: (888) REG-FAIR (734-3247)
- Fax: (202) 481-5719
- E-mail: ombudsman@sba.gov
- Mail: Office of the National Ombudsman U.S. Small Business Administration 409 3rd Street, S.W., MC2120 Washington, DC 20416-0005

To view the SBREFA Act in its entirety, please visit the following web link: www.sba.gov/advo/laws/sbrefa.html

For more information on SBREFA, the following web links may prove helpful:

www.sba.gov/ombudsman
www.sba.gov/ombudsman/dsp_overview.html
www.sba.gov/ombudsman/dsp_faq.html
www.sba.gov/advo www.sba.gov/advo/laws/is_oshapanel.html

NOTE: Filing a complaint with the SBA Ombudsman does not affect any obligation that you may have to comply with an OSHA citation or other enforcement action. Nor does it mean that you need not take other available legal steps to protect your interests.

INDEX

D

E

F

G

H

I

J

L

M

N

O

P

R

S

T

U

V

W